No. 1429
$14.95

PATENT IT YOURSELF

by Hrand M. Muncheryan,
B.Sc., E.E., M.Sc.

TAB BOOKS Inc.
BLUE RIDGE SUMMIT, PA. 17214

FIRST EDITION

THIRD PRINTING

Printed in the United States of America

Reproduction or publication of the content in any manner, without express permission of the publisher, is prohibited. No liability is assumed with respect to the use of the information herein.

Copyright © 1982 by TAB BOOKS Inc.

Library of Congress Cataloging in Publication Data

Muncheryan, Hrand M.
   Patent it yourself.

   Includes index.
   1. Patents—United States.   2. Trade-marks—United
States.   3. Copyright—United States.   I. Title.
T339.M924   608'.068   82-5696
ISBN 0-8306-2429-5   AACR2
ISBN 0-8306-1429-X (pbk.)

Contents

Introduction

This book shows you how to prepare a patent application for your inventions and how to successfully file an application in the patent office. With this book you can avoid the high cost of securing a patent through patent professionals.

This book will also aid those who are not attorneys-at-law, but who want to prepare themselves to become patent agents to practice before the patent office. For engineers and scientists, this book will be particularly useful in helping them analyze prior patents related to developmental work and avoid prior art in designing new devices and systems.

Included is information on how to make a search to uncover prior art, circumvent prior art, develop an idea into a working product, determine the patentability of a product, and successfully commercialize an invention. Other topics covered are utility patents, design patents, trademarks, trade names, and copyrights.

Each topic is discussed in a language easily understandable by the layman. There is a discussion on the component parts of a patent application and illustrations for each. There is an example of a complete patent application. How to prepare drawings in a manner consistent with conventional standards is explained.

Because the dominant part of a patent application resides in the claims—which define what the inventor considers to be his invention—special emphasis has been placed upon how to develop a claim that fully delineates an invention and that is broad in scope, yet difficult for others to circumvent. In addition, there is information on specific claims describing the specific parts of the invention in conjunction with broad claims; thereby the invention will constitute a superinvention.

Other sections of the book describe how as many as five species of the same invention can be included in a single application, how divisional application can be derived from a parent patent application, and how they fortify the parent patent. Additional sections take up the examiners' first action upon receiving an application. Outlined is the applicant's response and the examiners' action. Included is a comprehensive treatment of how to secure a trademark on the principal register or on the supplemental register (and their advantages and commercial merits).

To accomplish these tasks, you need only possess sufficient command of the English language to comprehensively communicate your ideas in a patent application and proceedings.

Chapter 1
Patent Office
Administration

United States patent laws were enacted by Congress in 1790. In 1802, the Patent and Trademark Office was established as a separate department headed by a Superintendent of Patents. By revision of the laws in 1836, the chief of the Patent and Trademark Office was designated as the Commissioner of Patents and Trademarks under the Department of State. In 1849, the Patent Office was transferred to the Department of the Interior. Congress made further revisions in the laws. In 1925, the Patent and Trademark Office was assigned a distinct department under the Department of Commerce (a branch of the federal government). The patent laws, as amended in 1975 and now in effect, are in Title 35. United States Code, Part I.

The chief functions of the Patent and Trademark Office are to administer the patent laws by granting letter patents for inventions and trademarks to individuals and companies having products designated with distinctive names. The office examines applications for patents and trademarks to determine whether the applicants are entitled to patents or trademarks under these laws. Upon a favorable determination of the content of an application, the office grants and issues patents to the respective applicants. After the issuance of a patent, the Patent and Trademark Office has no jurisdiction over enforcement of patents, questions of infringements, or related matters.

ADMINISTRATION OF THE PATENT AND TRADEMARK OFFICE

The administrator of the patent office is the commissioner of patents and trademarks (who is responsive to the secretary of commerce). The commissioner's staff comprises a deputy commissioner, several assistant commissioners of patents, and about 15 chief examiners. A patent solicitor's department is also under the commissioner. This department prepares and gives examinations to prospective patent attorneys who—upon passing a written examination and other qualifications set forth by the solicitor's office—become registered to practice patent law before the United States Patent and Trademark Office. The patent attorneys do not necessarily have to be lawyers in order to qualify for registration. Lawyers who pass the examination and become registered are known as *patent attorneys*; those that are not lawyers and are registered are called *patent agents*.

Under the direction of the secretary of commerce, the commissioner has the authority to conduct studies respecting the domestic and international patent and trademark laws. He may establish regulations, consistent with the law, for the conduct of Patent and Trademark Office proceedings. He also has supervision over the entire work of the Patent and Trademark Office. This includes prescribing rules for the conduct of the office proceedings, the recognition of patent attorneys and patent agents, and attending to the administration of the general activities of the Patent and Trademark Office.

The Patent and Trademark Office has 2700 employees. Half of them are patent examiners, and others have legal training and technical backgrounds covering many branches of the general technology. The office receives about 90,000 applications and over 3 million pieces of mail annually. Approximately 650 patents, together with a number of trademarks, are issued weekly. Abstracts of these patents with pertinent drawings are published weekly in the *Official Gazette*.

PATENT CLASSIFICATION AND LIBRARIES

Issued patents are classified by examiners so that examiners can readily and accurately determine the novelty of the invention for which applications for patents are on file. The examiners make use of the patent library for all the issued patents, trademarks, scientific books and periodicals (foreign and domestic) to aid the officers in the patent office in discharging their duties. In addition to the patent office library, duplicate patent libraries are located in the various states of the United States. An inventor who is in the process of developing an invention should visit the nearest library to make a search for a prior art. Such art, when found, not only might give him additional ideas for how to make an invention, but it also might reveal the prior art for the purpose of circumvention. The locations of these libraries are in Table 1-1. All matters related to the applications of patents and trademarks must be sent to:

The Commissioner of Patents and Trademarks
United States Department of Commerce
Patent and Trademark Office
Washington, DC 20231

PATENT OFFICE PUBLICATIONS

Publications and information on the following topics are available from the Patent and Trademark Office:

☐ Complete patents that include the headings of the drawings for purposes of photolithography.

☐ Certificates of trademark registration, drawings, and copies thereof.

☐ The *Official Gazette*, containing the United States patents and trademarks issued each week.

☐ Annual indexes of patents and names of the patentees, trademarks and the names of trademark registrants.

☐ Annual volumes of decisions in patent and trademark cases.

☐ Pamphlet copies of patent laws and rules of patent practice, laws relating to trademarks, and other publications related to patent office business.

☐ Printed copies of patents that are distributed to the public libraries.These can be procured by anyone at a cost of 50 cents per copy.

Because an inventor, by law, is entitled to prepare and submit his own patent application to the United States Patent Office, he should use the patent libraries (especially for prior art search) nearest his town. The patent librarian will be well versed in patent classifications and can aid the inventor in locating prior inventions and related matter. Other related material can be obtained by writing to the Superintendent of Documents, United States Printing Office, Washington, DC 20402.

PATENT FEES

The following fees are charged by the patent office for various publications and services:

Filing Fee. On filing an application for utility patent the fee is $150. In addition, $15 is charged for each independent claim in

Table 1-1. Patent Office Library Locations.

Alabama	— Birmingham Public Library
California	— Los Angeles Public Library
	Sacramento: California State Library
	Sunnyvale: Patent Information Clearing House *
Colorado	— Denver Public Library
Georgia	— Atlanta: Georgia Institute of Technology
Illinois	— Chicago Public Library
Massachusetts	— Boston Public Library
Michigan	— Detroit Public Library
Minnesota	— Minneapolis Public Library
Missouri	— Kansas City Public Library (Linda Hall Library)
	Saint Louis Public Library
Nebraska	— Lincoln: University of Nebraska
New Hampshire	— Durham: University of New Hampshire Library
New Jersey	— Newark Public Library
New York	— Albany: New York State Library
	Buffalo and Erie County Public Library
	New York Public Library
North Carolina	— Raleigh: North Carolina State University
Ohio	— Public Library of Cincinnati & Hamilson County
	Cleveland Public Library
	— Columbus: Ohio State University Libraries
	Toledo/Lucas County Public Library
Oklahoma	— Stillwater: Oklahoma State University Library
Pennsylvania	— Philadelphia: Franklin Institute Library
	Pittsburgh: Carnegie Library of Pittsburgh
	University Park: Pennsylvania State University Library
Rhode Island	— Providence Public Library
Tennessee	— Memphis and Shelby County Public Library
Texas	— Dallas Public Library
	Houston: The Rice University Fondren Library
Virginia	— U.S. Patent Office Library at Arlington
Washington	— Seattle: University of Washington Engineering Library
Wisconsin	— Madison: University of Wisconsin Library
	Milwaukee Public Library

* Collection Incomplete.

excess of one. There is a $5 charge for each claim in excess of 20 claims, whether independent or dependent.

Final Fee. There is a fee of $150.

Design Patent. On filing, the charge is $62.50 for each application. Final Fee is $67.50.

Reissue Patent. The charges are $250.

Revival of Abandoned Patent Application. The cost is $250.

Certification of Correction. The charge is $120 if the correction is due to the applicant's mistake; no charge if it is due to the patent office.

Disclaimer. On filing each disclaimer, send $15.

Revival of Abandoned Patent Application. The cost is $15.

Certification of Correction. The charge is $15 if the correction is due to the applicant's mistake; no charge if it is due to the patent office.

Recording Patent Assignment. The fee is $20. If two or more assignments are being recorded in the patent office, $3 is charged for each additional patent assignment.

All fees are to be paid to the patent office in advance (for a patent application or for a request for service). Payments may be made in money order or by personal check. Cash should not be sent. Any errors made in payment to the patent office can be rectified by the office by informing the applicant. United States Code, Title 35, Patents, Rule 42, states that the commissioner may refund any sum paid by mistake or in excess of the fee required by law.

PATENT MAINTENANCE FEE: PUBLIC LAW 96-517

Maintenance fees are to be paid to keep the patent in force during the 17-year period. The fees are to be paid as follows:

☐ First maintenance fee (3½ years after issuance of patent): $400.

☐ Second maintenance fee (7½ years after issuance of patent): $800.

☐ Third maintenance fee (11½ years after issuance of patent): $1,200.

☐ Failure to pay the maintenance fee according to schedule will cause the patent to become expired. Currently it is not clear how the public will be made aware of the expiration of the patent if the required fee has not been paid during the specified period.

Patent Applications Preserved in Secrecy (Section 1.14, CFR Title 37)

"(a) Except as provided in Section 1.11(b) pending patent applications are preserved in secrecy. No information will be given by the Office respecting the filing by any particular person of an application for a patent, the pendency of any particular case before it, or the subject matter of any particular application, nor will access be given to or copies furnished by any pending application or papers relating thereto, without written authority in that particular application from the applicant or his assignee or attorney or agent of "(b) Except as provided in Section 1.11(b) abandoned applications are likewise not open to public inspection, except that if an application referred to in a U.S. Patent, or in an application which is open to inspection pursuant to Section 1.139, is abandoned and is available, it may be inspected or copies obtained by any person on written request, without notice to the applicant. Abandoned applications may be destroyed after 20 years from their filing date, except those to which particular attention has been called and which have been marked for preservation. Abandoned applications will not be returned."

Section 1.139: Waiver of Patent Rights

"An applicant may waive his rights to an enforceable patent based on a pending patent application by filing in the Patent and Trademark Office a written waiver of patent rights, a consent to the publication of an abstract, an authorization to open the complete application to inspection by the general public, and a declaration of abandonment signed by the applicant and the assignee of record (if such exists)."

Chapter 2
Patentability
of Inventions

What is a Patent? A patent is a right or privilege issued, for a limited period, by the government to a person for his intellectual property (invention) so that he can transform said property into industrial or commercial value. In the United States, a patent grants the inventor the right to exclude others from making, using, or selling the patented invention for 17 years. The right granted is only to exclude others from exploiting the patented invention. Otherwise anyone would be free to make, use, or sell the invention.

Because inventions are primarily the products of mental faculty, as long as the inventor can keep his invention a secret he has control over it. There are shortcomings to this type of monopoly. Not all inventions can be kept secret because of their construction and method of assembly and operation. Someone else might invent a similar device. One of the chief purposes of acquiring a patent and having it published so that the public is aware of its construction, operation, and application is to promote the art of the particular invention. Technological progress becomes impeded without improvements in the manufacture, process, or method of making and contributing a useful product.

PROTECTION OFFERED BY A PATENT

A patent offers the patentee the right to prevent others from making, using, and selling a patented product. Any person who infringes upon the patentee's invention commits a trespass that constitutes a wrongful invasion of the inventor's property rights. A trespass is said to have been committed when an unauthorized person manufactures, uses, or sells the patented product without the consent of the patentee. The only protection a patentee has is the evidence that he is the inventor of the product, process, or method of manufacture by having his invention processed and registered in the patent office. The patent office makes a thorough search for anticipatory art prior to granting a patent. The only resort of the inventor is to sue the infringer in one of the United States District Courts. An injunction can be obtained against the sale of the invention by the infringer. In the event the patentee does not have the funds to bring a suit against the unauthorized party, the patent is a useless contrivance in protecting the patentee. In some European countries, the law is different and absolutely protects the inventor from having his patent rights violated by any infringer. The government of the country enters into the case with little or no cost to the patentee.

TYPES OF PATENTS

There are three types of patents recognized by the patent statutes. They are *utility patents, design patents,* and *plant patents.* In the language of the patent statutes, "a utility patent covers any new and useful process, machine, manufactured article, or composition of matter, or any new and useful improvement therefor." A design patent covers any new, original, and ornamental design for an article of manufacture. In a design patent, the appearance of the article must be new and original; in a utility patent, it is the usefulness of the mechanical, electronic, or chemical product that enters into the granting of a patent. In the case of plant patents, whomever discovers a method for asexually reproducing any distinct and new variety of plant, including *sports* (not spore), mutants, hybrids, and newly found seedlings is entitled to a patent. A tuborpropagated plant is not patentable.

PATENTABLE INVENTIONS

The patentability of an item depends on three requisites: invention, novelty, and utility. The mere idea of an article or process is not an invention. The idea must be first reduced to practice by making a model and operating it. Furthermore, the invention must not have previously existed. The parts for an invention may have existed for years and each part may have a distinct function during its existence. Nevertheless, when these parts are assembled together to produce an entirely new device of novel and useful function, the invention may be the subject of patentability.

8

Here are some typical questions and answers that define a patentable invention:

Can function (purpose, desired result) of a device be patented?

No. The purpose and application of an invention cannot be patented. However, any means for producing the function, if novel can be a subject of a patent.

Would it be a patentable invention to change the form or design of an existing invention?

No, if the change resides within the domain of mere construction. Yes, if the change brings about a new function or operation.

Are minor improvements within the range of expected skill patentable?

No. The invention must be the product of inventive facilities. It is difficult to convince the court that a minor improvement is in fact an invention. Also, the patent examiner will object to the improvement as being obvious and within one's skill.

Will it be patentable to change the chemical proportion of the ingredients of a chemical mixture to obtain a different kind of material?

Yes. For example, if in one proportion of rubber ingredients you obtain soft rubber and in another proportion you obtain a hard rubber, the properties are changed.

Would it be a patentable invention if superior and stronger material is substituted for a machine having weaker materials?

No, unless the superior materials change the function of the machine or it results in increased efficiency or increased saving in cost of operation.

Is it a patentable invention to apply an old invention to a new use?

No. The application of a device to a new use without changing its construction is not patentable.

Is combining an old device to a new combination patentable invention?

Usually no, unless the new combination brings about a new function or increased efficiency at a lower cost.

Is a newly discovered theory or law of nature patentable?

No. Nevertheless, a device to put the theory in operational form might be a patentable invention of the first magnitude.

Is a new process for fabricating an old material or device patentable?

Yes. For example, a new process for making nylon or for fabricating a safety razor would be a patentable invention.

Does the complexity of a device or system ensure patentability?

No, it doesn't if a simpler device or system does the same function just as efficiently.

Is a method of doing business patentable?

No.

Can a patent be granted for a mere idea?

No. The idea must be reduced to practice by constructing the device and operating it successfully. There must not be an existing prior patent.

Is increasing the parts of an old product to produce the same function a patentable invention?

Absolutely not.

In short, what constitutes a patentable invention?

A new invention may be a subject of a patent if it is included within the scope of the following characteristics:

☐ If the new product consists of a new construction and operation.

☐ Its new structure permits a new application for it.

☐ It shows increased efficiency over the existing products.

☐ The product uses parts that are lower cost than the earlier product with the same function.

☐ It is novel, useful, easier to use, and costs less than the old product.

☐ It uses fewer parts and yields greater efficiency than the older product.

☐ It is a new composition of matter with novel use.

CONDITIONS OF PATENTABILITY

The patent laws specify that any new process, machine, article of manufacture, composition of matter, or any useful improvement can be patented if they are novel, useful, and operable. Useless articles or processes are not patentable. No device that cannot operate would be a subject of patentability. Devices that are dangerous to handle and use are not patentable (an example is a weapon designed for committing murder). The following statutory conditions for patentability are specified on a patent application:

□ The applicant must have invented the invention.

□ The invention must not be known or used by others in this country, or patented or described in a printed publication before the invention thereof by an applicant.

□ The invention must not have been described in a publication in the United States or in a foreign country one year prior to the applicant's patent application.

□ The invention must not have been on sale or in public use in the United States or in a foreign country one year prior to the filing of the application by the applicant.

□ The inventor must not have abandoned the invention at any time prior to the filing of a patent application

□ The invention must not have been patented or a subject of an inventor's certificate filed in the Patent Office 12 months prior to the date of application of a patent by the applicant, his legal representatives or assigns.

□ The invention must not have been described and patented by another person in the United States or in another country one year prior to the applicant's application date.

□ A patent may not be obtained on a product if the difference of the product from that which is existing is so little and obvious that at the time the invention was made a person with ordinary skill could duplicate it easily.

□ If a prior invention exists, the applicant must have prima facie evidence to prove that he conceived prior to the existing invention and that he used diligence to improve his invention during the period prior to his patent application.

REDUCTION TO PRACTICE

An idea or a concept should be developed to a point whereby its utility and operation can be demonstrated before it can be considered to be an invention. To develop the idea, the first procedure is to make a drawing and description of the product. The description should enable an average person, skilled in the art, to understand and make a model from it. When a working model of the product is constructed and is operating in accordance with the requirement of the device, the result is an *invention*. In accordance with the patent statutes, the idea or concept has now been reduced to practice. When a patent application is filed, the idea or concept has been reduced to constructive practice. While it is not necessary to make a model before a patent application is filed, in the eyes of the law the invention is not reduced to practice before a working sample has been produced.

Constructing a working model preparatory to filing a patent application should be a requisite of the inventor. During construction of the first model, many problems that might present themselves can be corrected. After the first model is made, you should make an effort to improve it by using a more ingenious way of construction or by altering the design so that fewer parts are necessary to construct the invention. This procedure will improve the appearance, size, and possibly the efficiency of operation of the device. It will also reduce the cost of production. In addition, when a patent application is prepared, it will contain the best improvement the inventor can bring about by the time of filing the application. All improved mechanism and operational parts will be entered into the application. The claims will therefore be written on the best possible structure of the invention. After filing the application, if further improvements are made on the invention, the inventor may file an additional patent application to cover the new and further improved product (or the part thereof).

The patent office seldom requests the submission of a working model of an invention except in cases where the invention violates the laws of nature. An example of such an invention would be a machine for perpetual motion. To date, no person has been able to obtain a patent on a perpetual-motion device. Because the drawings accompanying the application usually give full information regarding the invention—and the models would take additional room for storing—the submission of models is generally not required.

Chapter 3
Preliminary
Patent Search

Preparatory to the preparation of a patent application, you should make a thorough search to determine the current state of the art of an invention. The sources of information would be (given in the order of importance) the issued patents, technical books, trade magazines, proceedings of scientific or technical societies, and the market in general.

SEARCH AT THE PATENT LIBRARY

The United States Patent and Trademark Office maintains a complete library containing over 120,000 volumes of scientific and technical books, the official journals of foreign patent offices, over 4 million United States patents and 8 million foreign patents in bound volumes. The patents are arranged in accordance with more than 300 subject classifications and 64,000 subclasses. Included among these is a complete set of the *Official Gazette* in numerical order. Naturally, it is impossible to examine all these books and patents. The patent librarian will aid the inventor in categorizing his invention and providing the inventor with most of the available information in the particular classification. The library is located at Crystal Plaza, 2021 Jefferson Davis Highway, Arlington, Virginia.

Library locations are listed in Table 1-1. Any inventor may examine his own case in any one of these libraries (with the aid of the patent librarian). This search will enable the inventor to determine whether his invention is patentable and if a strong patent can be obtained on the invention.

An inventor who has not done this preliminary investigation and has proceeded to construct his invention at some considerable cost might find it a very bitter experience when he discovers later that someone else has already patented "his" idea or invention. Millions of dollars are wasted annually because this first step of patent search is ignored by inventors or their representatives. Even after making a patent search and finding no prior art on the invention under consideration, this preliminary procedure does not guarantee that additional search by the patent examiner will not uncover other patents. Nevertheless, the risk of losing the invention because of the anticipating patents will be reduced to a minimum by first conducting a patentability search.

PATENT SEARCH THROUGH BOOKS AND MAGAZINES

Books and magazines in the public libraries are arranged in classifications. For instance, books on electronic devices and systems, chemical compositions, aeronautical subjects, mechanical devices, and the like are categorized on the shelves in their subject classifications. You could look up the subject matter in classified indexes and make a list of the books and magazines. You can then present the list to the attending librarian to obtain the particular publications. It might take more than one visit to the library, but it will be worth it to do so in the long run. Additional information can be found in product catalogs at the local library. These catalogs describe products that might be already on the market but have not attained success. The inventor can often save money by making his own search. His efforts probably will be more thorough than a Washington searcher to whom the inventor pays for the services rendered.

By studying the prior art and patents in the classification of the inventor's product, much valuable information can be gained by the inventor. For instance, suppose that an inventor has reduced his idea to practice, but he then has discovered that a certain part of the invention is not working in the manner he has planned. He can learn from the prior patents how others have done the same thing, or something similar to it, and perhaps he can correct the deficiency without infringing on the patentee's invention. In the event any one or more anticipating inventions are discovered by the inventor, then he could alter the design or parts making up the invention. He might even discover a simpler method of making his invention. Consequently, when he is ready for preparing the patent application, he

MARKET SEARCH FOR PRIOR PRODUCTS

Subsequent to a literature search in a public library, it will also be advisable to make a cursory search through different department stores, drug stores, hardware stores, and the like to determine if any product similar to your idea for an invention exists. If there is no other product, further search is not necessary. In the event a similar device is found during market search, the inventor could closely examine the device to see if it contains any similar features. If this cannot be immediately determined, it would be to your advantage to purchase a sample for diassembly and examination of the working parts of the device. This will give you a chance to see if any part of the product is in conflict with your invention. If so you can make an attempt to circumvent the mechanism or parts used. In this way, you will have a valid reason for filing a patent application and you will probably succeed in obtaining a patent.

The inventor may file a patent application without any search and let the patent examiner do the search. I have found this latter method convenient and successful because many of my inventions are generally basic rather than improvements on any existing invention. If any prior art is found during the examiner's search, then the inventor can abandon the application and file a new one with his improvement included. No new matter can be added to a pending application at the patent office.

ANALYZING PRIOR ART

Having collected all the information regarding prior art, the inventor can now analyze this information in the light of anticipation with his invention. A part that is similar to the inventor's product, but has a different function, does not constitute anticipation on the inventor's invention. *Analogous parts*, those parts that have different structures but similar functions, can be circumvented if the remaining parts of the product cooperatively produce the function intended. For example, if an inventor has an electric pencil sharpener and the prior art shows a manually operated pencil sharpener, both products having cutting edges, the inventor can circumvent the prior product, if patented, by claiming a gear for sharpening pencils.

If after analyzing the prior art, the inventor believes that his invention is fully superseded, then he should abandon that structure and attempt to develop a device that can produce the function intended in a different way than that indicated by the previous inventions. There are dozens of products on the market that per-

and attempt to develop a device that can produce the function intended in a different way than that indicated by the previous inventions. There are dozens of products on the market that perform the same function but are constructed differently. A function cannot be patented, but the means for bringing about the intended function may be a subject of patentability if prior art is avoided.

CIRCUMVENTION OF PRIOR ART

Upon careful examination and analysis of the various patents, products, and publications, if no prior art is found that will infringe on the inventor's product, the inventor's next step is to prepare a patent application. If any prior art is found, the inventor's next recourse is to circumvent the prior art; this is perfectly legal and permissible. The circumvention must be such that a broad claim, that cannot be circumvented by another person, can be drawn on the invention. The inventor must redesign the item, if necessary, and construct a device that is new in most respects so that it is not just a minor improvement over the existing product.

Some manufacturers, when approached by the owner of a patent who wants to sell his patent outright or on a royalty basis, procure copies of the patent and distribute them among their engineering staff to analyze the patent content to determine if circumvention is possible. If the analysis results in favor of the patentee, then the manufacturer will be ready "to talk business' with the inventor. If the invention can be readily circumvented, then the inventor's offer is turned down.

An inventor, can make his own circumvention of his first invention or patent and, if he finds other ways of making the same device, he can include in one application as many as five species of the same invention. Drawings of as many as five different ways of making the device can be included in his application. If the invention can be made, for example, in seven different ways, then the inventor may prepare a second application and include the other two species in the second application.

When you are ready to prepare a patent application include all analogous parts in the invention and refer to them "means to do so and so . . ." in broad claims. In other words, you must try to discover any weakness or incompleteness in your invention and complement it to prevent others from taking advantage of any weakness in the patent that will be issued.

When an inventor makes a thorough study of the prior art, existing market products, and has determined that competitive products fully anticipate (occur prior and are similar to his invention) on his invention, he must then decide whether his invention has sufficient merit from the standpoints of novelty, efficiency, and cost incentive. If he does not find sufficient novelty and advantage over the existing products, he should take no further action on it.

If he determines that his invention has sufficient novelty and advantage over the existing products and patents, then he should immediately proceed to prepare a patent application. He must retain all the papers describing his original invention, drawings, and improvements in a safe place and must not disclose them or their dates to any person not related to his invention. These papers will constitute prima facie evidence of his invention, originality, and the dates of conception, development, and diligence he has employed up to the time of filing the patent application.

Chapter 4
Parts of a
Patent Application

Only the inventor may apply for a patent for his invention. A patent issued to a person pretending to be the inventor will become void. A person falsely applying for the patent would be subject to criminal penalties by the government. Only when the inventor is deceased may his legal representative file an application for patent on an invention. In the event the inventor has become insane during the preparation of the models or application for a patent, his guardian may apply for a patent in lieu of the inventor.

A patent application is a written document directed to the commissioner of patents and trademarks. It consists of a petition, a specification of the invention, claims defining the scope of the invention and its novel features, an oath or declaration stating the applicant is the sole inventor and that no prior art exists in the category of the invention, and a drawing showing the various features that fully indicate the construction and assembly of the invention. For chemical patent applications, usually no drawing is necessary unless the constituent parts can be structurally represented to indicate the manner of attachment of the radicals to the main molecular structure. The application is signed by the inventor at several places as indicated in Figs. 4-1 through 4-7.

PETITION

The *petition* is a formal request directed to the commissioner of patents and trademarks for the issuance of a patent. It contains the full name of the inventor, his citizenship, resident address, city, state, and the zip code. In addition, it gives the title of the invention

```
PETITION BY A SOLE INVENTOR

To the Commissioner of Patents and Trademarks:

    Petitioner........................, a citizen of the United States and a
resident of.......... .........., State of..............., whose post-office
address is.............. ....., Pray that letters patent be granted to him
for the improvement in...................., set forth in the following
specification.                                Signed .............

                                         (Inventor's Full Signature)
```

Fig. 4-1. An example of a petition by a sole inventor.

for which a patent is requested. The petition could be placed separately on a single sheet or combined with the specification. The petition would be placed on top of the application file. It is signed by the inventor.

SPECIFICATION

The *specification* consists of a description of the invention. Reference is made to each of its parts with numerical characters labeled on the drawing. How they are interrelated and their operational functions are stated. The specification includes a brief abstract of the invention, a summary of the invention, a reference section to the figures in the drawing, and the main body of the specification that describes the invention in full (with reference numerals).

The specifications, beginning with the petition, is typed on legal-size typewriter paper having consecutive numerals vertically arranged on its left-hand side, from 1 through 32. The size of the paper is preferably 8½ by 13 inches. Each typed sheet is consecu-

```
PETITION BY JOINT INVENTORS

To the Commissioner of Patents and Trademarks:

    Your petitioners,....................and...................., citizens of
the United States and residents, respectively, of .................., State of
.............., and of..............., State of.................... whose
post-office addreses are, respectively,.................. and................,
pray that letters patent may be granted to them, as joint inventors, for the
improvement in...................., set forth in the following specification.

                              Signed...  ...............
                              (Full Signature of First Inventor)

                              Signed.....................
                              (Full Signature of Second Inventor)
```

Fig. 4-2. An example of a petition by joint inventors.

Fig. 4-3. An example of a petition for a patent by an administrator.

tively numbered so that reference can be made to them, when necessary, during prosecution of the application. The numerals on the left-hand side are separated by a distance equal to a double-space of the typewriter. Each typed line is thus separated one from the other by double spacing.

Abstract of the Disclosure. The specification begins with the title of the invention written in capital letters and centered at the top of the page on line 1. The phrase, ABSTRACT OF THE DISCLOSURE, is centered in capital letters on the next line after the specification. The abstract of the disclosure is a very brief description of the invention; it states what it is, how it operates, and what it does when operating. No reference to the drawing is made in the abstract. The length of the abstract should be kept under 100 words if possible. The abstract should not state any speculative applications the invention might have in addition to the main application described in the specification and claimed.

Background of the Invention. Following the preceding information, BACKGROUND OF THE INVENTION is given in concise and clear language. This section includes statements with reference to the conception of the invention, how it was brought about, and the need for such an invention and its development. In this section, several examples of the prior devices related to the present invention may be discussed. Emphasis may be placed on their shortcomings, failure to perform the function intended effectively, their high cost, size, configuration, and lack of utility (as appropriate). In this connection, statements may be made on how the present invention will overcome the shortcomings of prior inventions and how it functions more advantageously with respect to the competitive devices.

The statements may also include a reference to prior patents that the inventor has been granted, related to the present invention, and the modifications or improvements made on the present invention in comparative respect to the former inventions patented by him. If an applicant's prior invention is still pending, state the serial number, if a patent has been granted, state the patent number and date of issue. In this way, the present invention is clearly distinguished from prior patents.

Summary of the Invention. Following the preceding statements a SUMMARY OF THE INVENTION is given that describes

---

<u>OATH</u>

State of . . . .)
)ss:
County of . . . .)

. . . . the above-named applicant, being sworn (or affirmed), deposes and says that he is a citizen of the United States and a resident of . . . . . .
. . . . , State of . . . . , that he verily believes himself to be the original, first and sole inventor of the improvement in . . . . described and claimed in the foregoing specification; that he does not know and does not believe that the same was ever known or used before his invention thereof, or patented or described in any printed publication in any country before his invention thereof, or more than one year prior to this application, or in public use or on sale in the United States more than one year prior to the application; that said invention has not been patented or made a subject of an inventor's certificate in any country foreign to the United States on an application filed by him or his legal representatives or assigns more than twelve months prior to this application; that he acknowledges his duty to disclose information of which he is aware which is material to the examination of this application; and that no application for patent or inventor's certificate on said invention has been filed by him or his representatives or assigns in any country foreign to the United States.

                          Signed . . . .
                          (Inventor's Full Signature)
Sworn and subscribed before me this . . . . . day of . . . . , 19 . . . .
                          Signed . . . .
                          (Signature of Notary or Officer)
                          Signed . . . .
                          (Official Character or Seal)

Fig. 4-4. An example of an oath to accompany an application for a patent.

Fig. 4-5. An example of an oath by joint inventors (patent application).

the principal components. This summary may consist of short paragraphs describing the main functions of the components and their advantages. The advantages of the component might be its method of construction, reduction in the number of parts (with reference to prior art), reduced size and cost, higher efficiency than previous products, and flexibility of use with respect to the existing competitive products.

Brief Description of the Drawing. The title, BRIEF DE-SCRIPTION OF THE DRAWING, is placed and centered under the last line of the invention summary. A brief description of each figure in the drawing, by reference to the figure number, and the particular view of the drawing are given. For instance, it might refer to Figure 1 as being the perspective view of the invention, Figure 2 as being the cross-sectional view, and Figure 3 as being an enlarged sectional view of the invention, etc.

Detailed Description of the Invention. The last part of the specification consists of the DETAILED DESCRIPTION OF THE INVENTION. The title is centered on the line below the last line of the drawing description. This section begins by reference to the device shown in Figure 1, and refers to the main structure or body of the device as numeral 1, and continues by assigning consecutive numbers to different parts as they are described. Where applicable, the material, strength, or other characteristics of the parts are stated. Any alternative materials that can be used advantageously are also mentioned. As the description continues by reference to different parts, their interrelationship and functional characteristics are stated.

The description has a final paragraph stating additional applications for the invention and how the applicant may modify the invention to adapt it to the new applications *without* departing from the spirit and scope of the appended claims. This last paragraph does

---

DECLARATION

. . . ., the above-named petitioner, declares that he is a citizen of the United States and resident of . . . . that he verily believes himself to be the original, first and sole inventor of the improvement in . . . . described and claimed in the annexed specification; that he does not know and does not believe that the same was ever known or used before his invention thereof, or patented or described in any printed publication in any country before his invention thereof, or more than one year prior to this application, or in public use or on sale in the United States more than one year prior to this application; that said invention has not been patented in any country foreign to the United States on any application filed by him or his legal representatives or assigns more than twelve months prior to this application; that he acknowledges his duty to disclose information of which he is aware is material to the examination of this application, and that no application for patent on said invention has been filed by him or his representative or assigns in any country foreign to the United States.

The undersigned petitioner declares further that all statements made herein of his own knowledge are true and that all statements made on information and belief are believed to be true; and further that these statements were made with the knowledge that willful false statements and the like so made are punishable by fine or imprisonment, or both, under section 1001 of Title 18 of the United States Code and that such willful false statements may jeopardize the validity of the application or any patent issuing thereon.

Signed . . . .
(Inventor's Full Signature)

Date: . . . .

Fig. 4-6. An example of a declaration in lieu of an oath.

```
┌─────────────────────────────────────────────────────────────┐
│                          OATH                               │
│ State of . . . . . . . . . . . . . . )                      │
│                              ) ss:                          │
│ County of . . . . . . . . . . . . )                         │
│ . . . . . . . . . . . . . . . . . . , being sworn (or affirmed) deposes and says that he is a │
│ citizen of the United States of America and resident of . . . . . . . . . . . . . . . . . . . , that │
│ on . . . . . . , 19 . . . . . . . . . , he filed application for patent Serial No . . . . . . . . . in the │
│ United States Patent and Trademark Office, that he verily believes himself to be │
│ the original, first and sole inventor of the improvement in . . . . . . . . . . . . . . . . . . . , │
│ described and claimed in the specification of said application for patent; that he │
│ does not know and does not believe that the same was ever known or used │
│ before his invention thereof, or patented or described in any printed publication in │
│ any country before his invention thereof, or more than one year prior to the date │
│ of said application, or in public use or on sale in the United States for more than │
│ one year prior to the date of said application; that said invention has not been │
│ patented or made the subject of an inventor's certificate before the date of said │
│ application in any country foreign to the United States or an application filed by │
│ him or his legal representatives or assigns more than twelve months prior to the │
│ date of said application; that he acknowledges his duty to disclose information of │
│ which he is aware which is material to the examination of this application, and │
│ that no application for patent or inventor's certificate on said invention has been │
│ filed by him or his representatives or assigns in any country foreign to the United │
│ States.                                                     │
│                      Signed . . . . . . . . . . . . . . . . . .  │
│                      (Inventor's full signature)            │
│ Sworn to and subscribed before me this . . . . . . . . . . day of . . . . . . . . . . , 19 . . . . . │
│                      Signed . . . . . . . . . . . . . . . . . .  │
│                      (Signature of Notary of Officer)       │
│ (Seal)                                                      │
│                      . . . . . . . . . . . . . . . . . . .   │
│                      (Official Character)                   │
└─────────────────────────────────────────────────────────────┘
```

Fig. 4-7. An example of an oath.

not bear much significance except that it mentions that, if any new application for the invention exists, the inventor will be entitled to use the invention for that purpose. In any case, an inventor can use his invention for any application he desires without departing from the scope of the claims.

CLAIMS

The specification must conclude with one or more claims. The claims point out the subject matter that the inventor considers to be his invention. The number of claims that can be written on an invention is not limited by statutes. The novelty and complexity of the invention determine the number of claims that should be or can be written. For example, if an invention is such that practically no prior art exists, then to cover all and each of the novel parts entering into the construction of the invention could be claimed as a new invention. Likewise, if the invention has many parts to assemble it into a unitary structure and each part is sufficiently important in the operation of the invention, than as many claims as would be necessary to cover each part or a combination of parts working cooperatively must be covered by additional claims.

The two types of claims are *independent claims* and *dependent claims*. They are also known as *generic claims* and *specific claims*, respectively. An independent claim defines the scope of the inventor's monopoly covering the novel features of the invention. Every claim made by the inventor in pursuing the monopoly of his invention must be supported by sufficient disclosure in the specification and the drawings. If a claim is made on matter that does not exist in the specification, such a claim is declared invalid.

An independent claim is a broad claim and broadly covers all the elements of the invention. A dependent claim contains a reference to a claim previously set forth and explains the broadly defined terms in the previous claim to which it makes a reference. For example, if the independent or generic claim specifies "a means for illumination," the dependent claim may further limit it by stating "an electric lamp for illumination" or "a kerosene lamp for illumination." A means for illumination covers all types of light-emitting sources, but an electric lamp narrows the claim to only one type of source of illumination.

A claim may also be written by reference to a dependent claim by referring to its number and further defining the specific structure in that claim (such as "an electric fluorescent lamp"). If an independent claim states only "an electric fluorescent lamp," the patent covers only an electric fluorescent lamp. Anyone can make and patent a tungsten-filament lamp, a kerosene lamp, or a gas-mantle-type lamp without infringing on the electric fluorescent lamp patent, provided, of course, no prior patent on these lamps exist. The more prior patents exist the more limiting phrases are required in a claim to cover the improvement. A claim so narrowed down may become worthless because only a very small portion of the illuminating devices has been protected by a patent. This subject is treated more specifically in Chapter 5.

OATH OR DECLARATION

The *oath* declares that the inventor believes to be the first and original inventor of the article of manufacture, process, or compositiono of matter (or improvement thereof), and the inventor solicits a patent from the patent office. The oath states the citizenship of the applicant. The applicant declares that he has not seen a similar product or the publication thereof in any country one year prior to the filing of the application. It also states that the applicant has not filed an inventor's certificate in the patent office 12 months prior to the filing of his application. The oath is signed by the applicant before a notary public or an officer authorized by the government. The authorized person or the notary public signs the document and places his seal on the same page where the signatures appear.

A *declaration form* as prescribed by the commissioner of patents and trademarks can be made in lieu of the oath. In such an event, the document does not need the signature of a notary public or any authorized government officer. The declaration contains substantially all the declared matter in an oath. In addition another paragraph is appended. The paragraph states that willful false statement and the like are punishable by a fine or imprisonment, or both, in accordance with 18 U.S. C. 1001, which reads as follows:

"Whoever in any matter within the jurisdiction of any department or agency of the United States knowingly and willfully falsifies, conceals or covers up by any trick, scheme, or device a material fact, or makes any false, fictitious or fraudulant statement or entry, shall be fined not more than $10,000 or imprisoned not more than five years, or both."

DRAWING

When the nature of the invention admits of illustration, the applicant must submit a drawing of the invention together with the complete application. The drawing delineates the subject matter in such a way that a person skilled in the art can readily understand the construction and operation of the invention. As many features of the invention may be drawn, including cross-sectional or partial cutouts, to enable the reader to make and operate the invention when he is authorized by the patentee when the patent is still in force.

The drawing is made with india ink on a white two-ply or three-ply, 8½-by-14-inch Bristol board. The drawing is made on an area 2 inches below the top margin and a quarter inch from both left and right margins and the bottom margin. The drawing area then will be 8 by 11¾ inches. Each of the views of the drawing is made at a scale sufficiently large to permit the fine structures to be shown without crowding them.

The various parts of the drawing must be numbered consecutively as the description proceeds. The same parts that appear in several views of the drawing should be designated by the same number and labeled by the same name given in the description. The reference characters (numerals) should measure at least one-eighth of an inch in height so that when they are reduced during printing they will be about one-twenty-fourth of an inch. The reference characters should not be circled and they should not be placed in any complex part to interfere with the drawing of that section. Parts that cannot be identified graphically can be labeled by their names on the drawing. This type of referencing should be limited to only a few such names per sheet of drawing.

INVENTOR'S SIGNATURE

The inventor must sign the petition, specification (after the claims), and the oath before filing the patent application. At the left-hand side, across the inventor's signature, the date and the inventor's residence (city and state) should be indicated. The signing of the drawing by the inventor is not permitted. For purposes of identification, the name and the address of the inventor may be placed and centered at the very top margin in a space one-half of an inch wide and 2 inches long.

When a product is invented jointly by two or more persons, they must apply jointly and each must sign the application and the required oath or declaration. In the event a joint inventor refuses to join in an application for patent or cannot be found after a diligent effort, the application can be made by the other inventor in behalf of himself and the omitted inventor. The commissioner should be made aware of the existing facts regarding the missing joint inventor or of his refusal to sign.

PATENT FILING AND FINAL FEES

The application for filing must be accompanied by the prescribed fee of $65, plus $10 for each independent claim in excess of one, and $2 for each dependent claim in excess of 10 claims. Any error in payment can be rectified by the patent office. The office sends a notice to the inventor and asks for the additional amount needed or returns the amount paid in excess. If the error occurs in the filing fee, this condition does not affect the filing date.

When the patent application is allowed, the inventor receives a notice for payment of the final fee of $100, plus $10 for each printed page and $2 for each drawing sheet. The final fee must be paid within three months after notification of the allowance in order to avoid lapsing of the application. If the inventor wants to extend the time of payment of the final fee, he may write to the commissioner for an extension to six months.

Chapter 5
Preparation
of Patent Claims

Regardless what is described in the specification and shown in the drawings, a patent dominates only what is claimed in the final form of the application. While the specification might refer to other prior inventions and distinguishes the improvement and the advantages of the present invention over the prior art, the claims define only the scope of the patentee's monopoly of his invention. The claims must be worded to cover what the inventor believes to be his invention and nothing more. A claim is dependent on what has been stated in the specification and shown in the drawing. All the referenced names of the parts in the specification must correspond and be the same in the claims. For example, a specification may be written so that it describes a mirror as a reflector, a reflecting surface, or a reflecting element. The claim may use any one of these names. In addition, to broaden the scope of the claim it may state "a reflecting means," that includes all types of reflectors or mirrors. Always remember that a patent is as good and strong as the claim or claims that define it.

TYPES OF CLAIMS

Claims are classified as generic or independent claims and specific or dependent claims. To write a good, broad claim, the inventor must have a high degree of knowledge about his invention and about the novelty as distinguished from prior art. In writing the claims, the inventor must remember to avoid the prior art in the claim. Frequently when an inventor or his representative files a patent application it is discovered upon first office action that the invention has been superseded by prior patents. For this reason, the preliminary search and the thorough examination of prior art cannot be overemphasized. A good claim is a good patent in itself covering the whole or a part of the invention. The strength and broadness of the scope of a claim is dependent on the nature of the prior art. With few anticipating disclosures, the inventor's chances of drafting strong claims to cover the invention is greater than when the field is very competitive with prior products that are similar in function and application.

Generic or Independent Claims

A generic or independent claim has a broad construction and contains all the important elements and limitations of the invention. It is usually written shorter with broader scope and use of terminology. For example, a claim on a typewriter with no prior art may be written as "a typewriter having a housing provided therein with means adapted to produce characters on paper when mounted thereon." This is considered to be a broad claim because the statement does not specify whether the typewriter is manual or electric. It does not specify whether the "means adapted to produce characters" is a typewriter key having letters, numerals, or linotyping. Also, it does not state whether the typescript is in English language or any other language. The language of the claim covers all types of typewriters, all types of typing elements, and all types of letters, numerals, or other symbols. "Characters" covers all the elements used in a typewriter.

Specific Claims

A specific claim is recited on a typewriter claim when the field is crowded with different types of typewriters because the invention cannot be covered with a broad claim. For example, a specific claim on such a typewriter is "a manual (or electric) typewriter having a housing provided therein with typewriter keys having English letters for printing words on typewriter paper when mounted thereon." This claim is directed to a manual (electric) typewriter with English typescript. This claim can be circumvented because the typewriter can be made to operate electrically (if the word manual is used). The keys could be made with a set of foreign alphabets, numerals, other symbols—or all of them. The broader claim is made with fewer words and broad definitions. The narrower claim contains more words with specific meanings.

Let's draft a claim for a spoon. If no patent on a spoon exists, a generic claim would be "a spoon." If a patent on a wooden spoon had previously been granted, then a specific claim could be written as "a metal spoon." If patents on both the wooden spoon and the metal spoon exist, then the claim has to be narrowed down something like "a metal spoon having a brass core plated with silver." It is readily apparent that if a patent is granted for "a spoon" it will dominate all types of spoons whether they are made of wood, plastic, metal, or glass. Whether they are made of any type of metal plated with silver, gold, nickel, or platinum will be dominated by one generic claim.

After one or more generic claims are drafted, such as "a spoon having a core with cladding," it will be advantageous to claim additional specific claims if they have predecessors. For instance, if the specification states that the spoon is made of any suitable core plated with noble metals, such as silver and gold, the generic claim is immediately understood to include a spoon plated with any one of the noble metals. The specific claim will actually recite the names of the plating metals.

Let's write a few broad and specific claims covering a desk lamp.

Claim 1. A desk lamp comprising a base provided with an elongated means with one end supported on said base and the opposite end provided with an illuminating means.

Claim 2. A desk lamp comprising an oblong base with a flexible metal gooseneck, with one end welded to said base and the other end having a socket for holding an electric light bulb.

Claim 1 is considered to be a broad or generic claim because it covers all types of desk lamps that give off illumination. Claim 2 is a narrow claim and can be circumvented because, instead of an oblong base, a square or circular base can be used. And instead of a flexible metal gooseneck, an inflexible tubular plastic structure can be used. The opposite and free end of the tubular structure could be provided with a quadrangular shade or reflector having a fluorescent tube extending along the inside cavity of the shade. Accordingly, it should be clear that an invention is dominated only by what it states in its claims. Any deviation from what is stated can be a subject of patentability if any prior art does not exist.

A broad or generic claim is shorter than the specific or narrow claim. This is true when the article to be patented is a simple device and when there are few prior patents. In the case of more complex inventions, where a plurality of parts are involved in the construction of the devices, short and broad claims cannot be drafted. The language of the claim may refer to different parts as first means, second means, third means, and the like. Because the term *means* covers a number of analogous parts, the claim is still a broad claim and it also could be generic if several species are included in the drawing. Such a claim is also an independent claim because it does not depend on any other claim. Other claims may be written dependent on this first claim by reference to it as the principal device and specifying what the first means, second means, etc. stand for. In such an event, the main broad claim can be long and the specific claims can be written briefly and dependent on the first claim.

DRAFTING A PATENT CLAIM

A patent claim defines the invention the applicant regards to be his original work and is novel with respect to prior art. The claim is included in a single paragraph containing a number of clauses or phrases ending in commas or semicolons. The last clause or phrase in the claim terminates with a period. Applicants must keep in mind that every claim may be considered as a whole patent. From that standpoint, there are a few generalized rules to follow.

☐ In describing the invention, the claim must detail an operative structure. A metal edge with no function cannot be patented. When the edge is sharpened so that it can cut as a knife, it has a definite function and utility. Therefore it can be patented if no prior art exists.

☐ A claim must recite the part in an invention in cooperative relation with another part or parts with which it is connected. A recently invented eraser provided at one end of a ball-point pen cannot subject the entire pen to patentability as a new device because there is no interrelationship between the writing end and the eraser. Each can be used independently of the other. If there is novelty in the material of the eraser, then a patent may be granted thereon.

☐ A claim must specify the principal elements contained in the invention. For example, instead of stating "a desk lamp illuminating the top of the desk," you should state "a desk lamp having a source of light to illuminate the top of the desk." Here the claim specifies the source of light; without it the desk top cannot be illuminated.

☐ In defining an invention which has an aperture, the claim should not set forth the aperture as the principal structure, but it should recite the part which contains the aperture. For example, you should state: "a circular window plate having an aperture in the center thereof . . ." rather than "an aperture located in the center of a circular plate . . . ." The aperture is not a part. No claim can be drawn on it.

☐ A claim should not be written as an aggregation of parts put together to form the whole structure. Such a claim will always be rejected by the examiner. For example, "A chair having legs, a back, and a seat for sitting thereon . . ." does not indicate any cooperative function of the different parts. But, "A chair having a seat provided with means for supporting said seat a suitable distance from the floor, and a back attached to said seat and extending upwardly therefrom," is a patentable invention when it is a novel piece of furniture.

☐ A claim should not contain any duplication of parts therein without appending a new part to the duplicated statement. For example, "a chair having a seat provided with means for supporting said seat a suitable distance from the floor, said seat having a back extending upwardly therefrom, and two side arms attached to said seat and said back." Two arms are added to the previous claim. This makes it a new invention as well as a new claim.

☐ For defining the different parts of an invention, a claim should not refer to the reference characters in the drawing. When the patent office first started operation, such a practice was common. The rule has been abandoned.

☐ A claim should not define the different elements alternatively. For example, "The wing structure can be made of aluminum, magnesium, or titanium." It should state, "The wing structure is made of a nonferrous metal characterized by aluminum, magnesium, and titanium."

☐ A claim must not cite matter in a prior patent whether it belongs to the same inventor or not. The claim, when thus recited, is declared to be *double-patenting* the invention and could damage the scope of the new patent to the extent of invalidating it.

☐ When a part in a claim is difficult to describe because of its particular constituents or shape, the claim may contain a negative term. For example, if an alloy contains beryllium and copper, it may be stated as a "nonferrous material." An odd-shaped part may be stated as a "noncircular plate."

☐ If more than one dependent claim is given in a patent application, each claim should be written patentably different from the other. For example, note the claims in patent No. 4,180,810.

☐ In a claim, no statement should be made that is not necessary to bring about the desired function in a device. For example, ". . . said element is provided with a sharp cutting edge capable of drilling holes in stainless steel, tungsten, and titanium, during which operation dense fumes of metal are noted being emitted." The emission of fumes is not a function of the drilling element and, therefore, should not be stated. If a fluid is used in drilling, no fumes will be emitted.

☐ The terms and phrases used in the claim must have antecedent basis in the specification. They must specify only essential features to differentiate the invention from prior art.

DRAFTING CLAIMS WITH IMAGINATION

In drafting claims, an important task is to write a broad claim that not only covers the present invention, but also covers any improvement on the item that *can* be made either by the inventor or a competitor after it is on the market. If an inventor, while preparing a broad claim, uses imagination to include variations of the invention he will cover extensive ground in the field of his inventions so that circumvention of his invention will be very difficult.

The drafting of the claim also involves logic and proper selection of words that could include additional variations expected in the future. For example, if the invention is on an unpatented electric light bulb, instead of stating "an electric, tungsten-filament light bulb" or "an electric fluorescent tube,", you could cover lamps and additional variations by writing "an electric light-emitting means." Such a light-emitting means can be any type of lamp electrically energized to light emission. For example, it could also be a neon light, a television screen, a laser emitter, and even a semiconductor light-emitting diode that emits light when stimulated by an electric current.

Drafting of a claim, whether it is broad or narrow, requires a manipulation of language defining the invention in its broadest, clearest, best understood, and definite terms. On the other hand, if the language of a broad claim is too broad—such as "a light emitter," that includes the sun and the stars—the claim will be useless. Certainly, the inventor does not mean to include such claims. Because the claims depend upon what is stated in the specification, a claim containing the words "light emitter" would be invalid because it defines something too abstract and it has no antecedent in the specification. Electric light emitter will be broad, but it would define clearly that the light is produced as a result of electric current passing through the invention. This action would be further defined in the same broad claim by stating the principal elements that cooperatively take part in bringing about the production of light by the passage of current through them.

SPECIES CLAIMS IN AN APPLICATION

The word *species* as applied to the description of invention relates to a group of modified forms of the invention. When an invention can be made in more than one way (for example, three ways), the official drawing may exhibit two more drawings in addition to the principal invention. In an application of this type, three species exist each is the modification of the other two. In accordance with the U.S. Patent Rule 1.141, two or more independent and distinct inventions may not be claimed in one application, with the exception that more than one species of an invention, not to exceed five, may be specifically claimed in the same application. This is on condition that the application also contains a generic claim that covers all the species.

Let's assume that a certain invention consists of seven species. Five of these species can be included in one application and the other two in a second application. In order to take advantage of the filing date, all seven species can be included in one application. In such an event, after the first office action, any two of the species can be removed from the original application and a divisional application using the original drawing or a part of the drawing containing the two species can be used. Naturally, a new filing fee—together with an entirely new application—must be prepared with all its component parts for filing. If two or more independent and distinct inventions are included in a single application, the patent examiner will require the applicant to separate them and file a divisional application for each independent and distinct invention. The divisional application can be made any time before the final action has been taken.

A patent (No. 2,866,338), granted to me contains five species of an invention entitled Temperature-Indicating Device and Closure Cap and illustrates how different species of an invention can be drawn and how a generic claim may be prepared to cover all of them. This is in spite of the prior art containing seven United States patents and one British patent cited against the application during processing. It is apparent that the generic claim covering all the limitations of the species has circumvented all other prior patents.

It is said that an allowable generic claim is one of the most difficult tasks that confronts a patent writer. Because of the claim's broad scope, it requires a high skill in the art to avoid prior patents. In the event an allowable generic claim cannot be drafted because of the crowded prior art, a new application for each species of the invention must be filed. Each application will bear the filing date of the original application.

A generic claim to cover the five species, such as those shown in Fig. 5-1, can be presented as follows:

Claim 1. A temperature-indicating device and closure cap, comprising an annulus adapted to be secured on a nursing milk bottle and having an internal flange, a cap member having a dome with a magnifying window therein, means securing said cap member to the internal flange of said annulus and forming a chamber in said cap, an indicating dial with temperature-defining means thereon transversely positioned within said chamber in adjacent relation to said window, and a thermosensitive means pivoted within said chamber and attached to said dial with temperature-defining means for movement thereof with reference to said window under changes of temperature within said chamber.

This claim has been allowed with 25 other dependent and independent claims. While the field of a temperature-indicating device is crowded with prior patents, the new invention has the merits of novelty and utility. It meets the requirements of the patent statutes.

Analyzing this claim with respect to the five species included in the drawing, the analogous parts contained in each species can be listed as follows although they are structurally different.

☐ A closure cap with an aperture (all species have it).

☐ A cap member having a chamber and a magnifying window in its dome (the claim does not mention where the window is, and all species have a window).

☐ Means (peripheral flange) securing said cap member to said annulus (all species are arranged the same way).

☐ An indicating dial with temperature-defining means (all species have it).

☐ Dial positioned transversely in adjacent relation to said window (all species have the same arrangement).

☐ A thermosensitive means pivoted within said chamber (whether in center or radially, all species have the same arrangement).

☐ The thermosensitive means is attached to said dial (in all species).

☐ The dial moves by said thermosensitive means (all species are the same).

☐ The dial moves with reference to said window by change of temperature within said chamber (all species have the same arrangement).

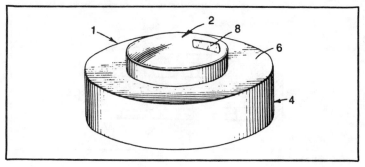

Fig. 5-1. A perspective view of the device.

All five species contain every part, function, and characteristic as defined in the generic claim. If the general claim failed to specify any one of the above listed items, such as the thermosensitive element, then the claim is not complete and its generic characteristics do not exist. Possibly such a claim will not be allowed by the patent examiner because the invention is inoperative. An inoperative invention cannot be patented.

Other claims dependent on this (generic) claim may be written detailing specific names referring to the broad claim. For example, the temperature-defining means can be identified as a dial having colored areas; each would define a different range of temperature as described in the specification. The phrase "means securing said cap member to said internal flange" refers to a peripheral flange provided on the cap member, and so on.

Prior to starting to write a generic claim, you must list all the analogous parts contained in different species and specify your claim on parts cooperatively and successively being interlinked in the construction of the device. Then you must broaden the language of the claim by eliminating the part name by using a general term such as *means*. If you are writing about the "colored areas," you can refer to them as temperature-defining areas because, if the areas are not colored, they may be shaded or etched differently with each shade representing a range of temperature.

If no generic claim can be prepared for an application, a divisional application must be filed bearing the filing date of the original application. The original application will then state in its preamble, before it becomes final, "This application was divided into a divisional application, now Serial No. ...." The divisional application will contain on its preamble, "This application is a continuation

37

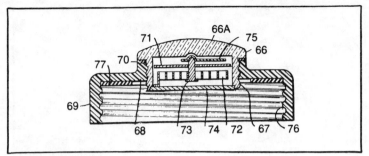

Fig. 5-2. Species No. 1.

of the application filed on .... and Serial No. ...." If a
patent has been issued on the original application, then the divisional application should state, "This application is a continuation of
the application filed on ...., now a patent, No. ...." In
this manner, the original and the divisional applications will be tied
to each other and properly identified.

A dependent claim (species claim) on the generic claim of the
temperature indicating-device will be as follows.

Claim 2. A temperature-indicating device as defined in Claim
1, wherein said temperature-defining means on said dial is a plurality of colored areas characterized by blue, green, and red, respectively representing cold, normal, and hot temperatures.

Claim 3. A temperature-indicating device as defined in Claim
1, wherein said thermosensitive means pivoted within said chamber
and attached to said dial therein is a coiled thermostatic ribbon-like
element rotative with said dial by responding to the changes of
ambient temperature therein.

Claim 4. A temperature-indicating device as defined in Claim
2, wherein said dial containing thereon blue, green, and red areas to
represent respectively cold, normal, and hot temperature, is provided with an L-shaped section thereunder, one end of said

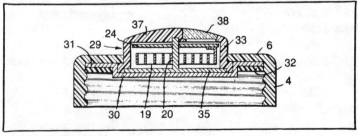

Fig. 5-3. Species No. 2.

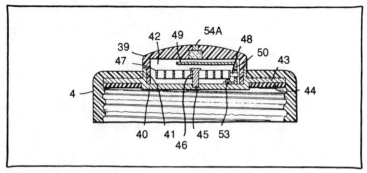

Fig. 5-4. Species No. 3.

L-shaped section being connected to one end of the thermosensitive element; a pin centrally located in said chamber and connected to the opposite end of said thermosensitive element, which, when activated by change of temperature therein, rotates said dial with temperature-defining areas thereon.

The original patent of the temperature-indicating device contains 26 claims. The reader must try to form additional dependent claims as he analyzes the drawings shown in Figs. 5-1 through 5-8. For example, try to draft claims for the following structures.

☐ The annular housing enclosing the temperature-indicating device.

☐ The magnifying window. In one claim it can be located in the center of the dome of the cap. In another claim it can be located radially to the dome.

☐ In Fig. 5-3, a pivot is located at the center of the flanged member and is connected to it.

☐ In Fig. 5-2, the temperature-defining dial is not circular and it is connected to the center of the thermosensitive element to be rotated thereby.

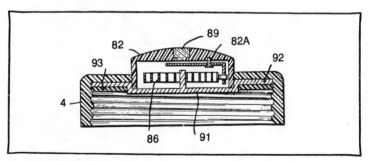

Fig. 5-5. Species No. 4.

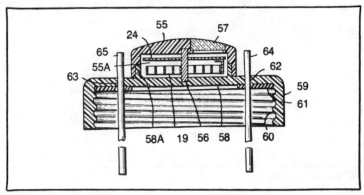

Fig. 5-6. Species No. 5.

☐ In Fig. 5-2, the peripheral end of the thermosensitive element is connected to the bottom disk of the temperature-indicating device, and its opposite end is connected to the temperature-defining dial.

☐ In Fig. 5-4, the temperature-defining dial has a projection bent downward and connected to the peripheral end of the thermosensitive element. The central end of the thermosensitive element is connected to the central pivot provided on the flanged bottom of the temperature-indicating cap.

☐ In Fig. 5-6, two tubes are projecting through the cap member and into the bottle containing the liquid. One of these tubes

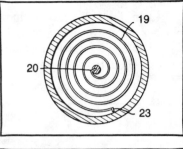

Fig. 5-7. A thermosensitive element.

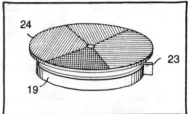

Fig. 5-8. Temperature-indicating colored areas.

is used for suction of the liquid from the bottle and the other is used to let air enter the bottle for easy flow of the liquid, such as a medical solution.

A dozen more dependent claims can be written by further study of the diagrams (Figs. 5-1 through 5-8). You should locate these structures and write claims as guided by claims 2, 3, and 4.

Chapter 6
Filing a
Patent Application

After a patent search is made and an inventor is reasonably confident that a strong patent can be obtained on his invention, he should prepare an application and immediately file it. The application must include a petition, an abstract of the disclosure, the specification, the claims, and the oath or a declaration. A complete patent application is given in this chapter. Start by preparing each part sequentially.

As an example, let's prepare a patent application on a silent awakening system. The system, shown in Fig. 6-1, is a device for awakening deaf persons as well as persons with normal hearing. It operates by the sound of a fire-detector alarm as well as by means of a clock to which it is connected electrically. The clock is set to an awakening time, when the time arrives, the clock triggers the awakening module placed under the pillow of the sleeping person. Persons who want to get up early in the morning, without disturbing other household members, can use the module advantageously.

The application is typed in the sequence and format shown in this chapter. The claims continue as shown in the printed patent until all pertinent claims have been entered. At the end of the last claim, the specification is signed by the applicant as follows:

January 16, 1978
Orange, California

_____

Hrand M. Muncheryan, applicant

Following the claims section, an oath or declaration is prepared on a separate sheet in accordance with one of the forms which

Fig. 6-1. A silent awakening system.

is applicable to the present application. For instance, if the applicant is a sole inventor, the oath related to "sole inventor" should be used. If the application is by joint inventors, the oath related to the "joint inventor" should be used. Both inventors should sign the oath before the notary public. If the inventor chose to use a declaration form, then he should sign the declaration and date it (without the necessity of a notary public).

The entire application consisting of the petition (Fig. 6-2), specification, claims, and the oath should be stapled together backed by a legal-size (blue) jacket to form a complete file. In mailing the file to the patent office, the drawing should be included loose in this file. In making a package of the file, two corrugated pieces of cardboard of the size of the drawing, should be used to "sandwich" the file. The package should be labeled DO NOT FOLD

PATENT APPLICATION
IN THE UNITED STATES PATENT AND TRADEMARK OFFICE

PETITION

To the Commissioner of Patents and Trademarks:

Your petitioner, Hrand M. Muncheryan, a citizen of the United States and a resident of the city of Orange, state of California, whose post office address is 1735 N. Morningside Street, Orange, California 92667, prays that letters patent be granted to him for improvement in SILENT AWAKENING SYSTEM, set forth in the following specification.

(Signed) _____

Hrand M. Muncheryan, Applicant

Fig. 6-2. An example of a petition.

so that the carrier will take special care not to damage it during transportation.

The entire patent application, Patent No. 4,180,810, follows. Note that the printed patent is slightly modified, but it includes the entire information sent to the patent office by the applicant. The applicant must draw a single drawing that represents the entire invention, if possible, so that it can be printed in the *Official Gazette*.

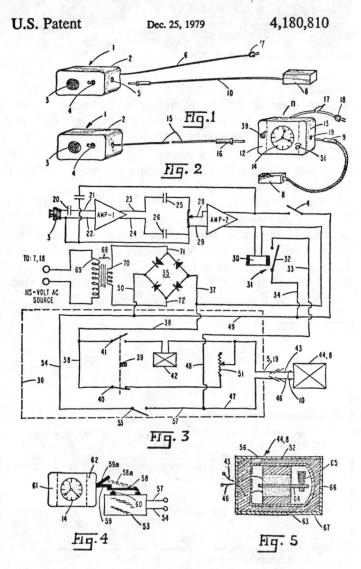

United States Patent [19]

Muncheryan

[11] **4,180,810**

[45] **Dec. 25, 1979**

[54] SILENT AWAKENING SYSTEM

[76] Inventor: Hrand M. Muncheryan, 1735 N. Morningside St., Orange, Calif. 92667

[21] Appl. No.: 869,797

[22] Filed: Jan. 16, 1978

[51] Int. Cl.² G08B 1/08
[52] U.S. Cl. 340/407; 340/148; 340/309.1
[58] Field of Search .... 340/309.1, 407, 148; 58/152 B

[56] References Cited

U.S. PATENT DOCUMENTS

3,786,628	1/1974	Fossard	340/407
4,028,882	1/1977	Muncheryan	340/407

Primary Examiner—Harold I. Pitts

[57] **ABSTRACT**

A silent awakening system operable either by receiving a sound-alarm signal, such as that from a fire-detection alarm, or by the actuation thereof by means of an alarm-triggering means of an electric clock, with the sound-producing mechanism of said clock being deactivated. The system is used to awaken deaf persons as well as persons of normal hearing, without disturbing others sleeping in the same room or nearby. It comprises a sound-receiving means which converts said sound to an electric signal to operate an electric relay, an electrical signal-processing circuit connected to said relay and actuated thereby, an electric alarm clock to set the time of awakening with the alarm-triggering means thereof mechanically connected to a switch means disposed in said electrical signal-processing circuit, and an awakening module electrically connected to said electrical signal-processing circuit to receive, for operation thereof, a processed current either through said electric relay or through said switch means actuated by the alarm-triggering means of said clock.

13 Claims, 5 Drawing Figures

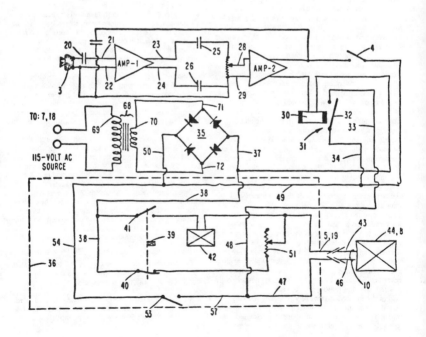

45

SILENT AWAKENING SYSTEM

The present invention is generally related to a device for awakening a sleeping person and is more particularly concerned with a device adapted to operate either by the sound of a fire-detector alarm or by the actuation of the alarm-triggering means of an electric clock set to a predetermined time of the day for awakening.

BACKGROUND OF THE INVENTION

At the present time, the general method of awakening of a person at a predetermined time is the use of an electric alarm clock, which triggers a sound alarm at a time set for awakening; this is accomplished by the stimulation of a person's auditory mechanism. However, if that person's auditory mechanism is not functioning properly, he cannot hear the sound and cannot be awakened by such means. One other method is the use of a flashing light operable by a clock set to actuate the flashing of the light at a desired time of awakening. The disadvantage of the former method is that the setting off of the alarm sound by the clock awakens other persons sleeping in the same or nearby room. The disadvantage of the latter method is that the sleeping person must be facing the flashing light at about the time of awakening; if the sleeping person is lying on his back or on his side away from the flashing light direction, the awakening action of the flashing light will be ineffective. Furthermore, the flashing light may also awaken other persons facing the flashing light.

To overcome these shortcomings of the existing methods, the present invention is developed to produce an effective and physical stimulating signal to awaken both deaf persons and persons of normal hearing, regardless of the position in which they are sleeping in bed. The present method affects only the person using the device and no one else. One such device is the applicant's prior invention using radio waves emanating from a means disposed either in a fire-detector system or in a clock, either of which could be located hundreds of feet away from the awakening unit, which is either placed under a bed pillow at night during sleeping or in a person's pocket during the day while working remotely from the alarm signal produced by either the fire detector or the clock. This invention is described and claimed in the applicant's patent application, Ser. No. 769,344, filed Feb. 16, 1977, with all the claims allowed and now pending for the issuance of a patent. In contrast with the operative characteristics of that device, the present invention employs audio-frequency signals emanating from a smoke-fire detector or other sounds to energize the awakening module, the signal-processing circuit thereof being remotely located in the sound detector of the system or in the electric clock, so that either the sound detector or the clock can be used independently of each other or jointly (by incident) to produce a silent awakening stimulus in the awakening module to awaken the sleeping person. Furthermore, the cost of manufacture of a system of this type is relatively lower than the radio-wave operated device and is within the financial means of an average working person, deaf persons, senior citizens, or others of low income. The system is also small and compact and easily portable from one place to another when desired.

SUMMARY OF THE INVENTION

The present invention is a system to silently alert a person to the dangers of toxic gases, smoke, fire alarm and other sounds which are particularly inaudible to deaf or partially deaf persons. The invention is also provided with means connected to and actuated by the alarm mechanism of an electric clock to produce electric signals which are transmitted to an awakening device placed under the pillow of a sleeping person to transmit thereto undulatory pulsations for awakening him. To achieve this purpose, the invention is provided with an electric circuit therein adapted to receive an acoustic signal, to convert said acoustic signal into an electric signal, and then to amplify it. The amplified output from this circuit is fed to an electric relay switch means connected in the electric circuit of said awakening device for energization thereof.

Another object of the invention is the provision of an awakening module which is connected to said electric relay switch means through an electric current source, which is provided, in the circuit thereof, with a current interrupter to produce a pulsative current flow through said awakening module.

A further object of the invention is the provision in said awakening module of a rotating means having a shaft with a weight attached to one terminal portion thereof and eccentrically mounted thereon for producing an unbalanced shaft rotation in said awakening module and thereby to create an undulatory or vibratory motion in said awakening module.

A still further object of the invention is to convert the undulatory or vibratory motion in said awakening module into pulsative undulations by the interrupted current fed into the rotating means of said awakening module, whereby a pulsative, undulatory motion is transmitted to the sleeping person through the pillow under which said awakening module is normally located for effective, stimulative operation thereof.

Another object of the invention is the provision of a means in said awakening module whereby said module can be connected either to the sound receiving and amplifying circuit of the system or to the alarm-producing mechanism of an electric clock to trigger a current flow into the awakening module circuit for awakening operation thereof.

A further object of the invention is to utilize any 115-volt 60-cycle household current for the operation of the awakening system, after said household current is properly processed in the system circuit by a means adapted to convert said household current into a current of a character utilizable by said awakening module.

One other object of the invention is the provision in the electric circuit thereof means to channel the current in said electric circuit into two channels, one channel passing therethrough a continuous current and the other producing an interrupted current therein prior to passing said current therethrough.

BRIEF DESCRIPTION OF THE DRAWINGS

Other objects and advantages of the invention will become more apparent from the specification taken in conjunction with the accompanying drawings, in which:

FIG. 1 is a plan view of the sound-operated silent alerting device together with an awakening module detachably connected thereto, forming the silent awakening system.

FIG. 2 is a plan view of a modified embodiment of the invention wherein both the sound-operated device and the awakening module are electrically connected through an electric alarm clock, whereby said awakening module can be operated either by the sound-operated alerting device or by the triggering of the alarm mechanism of the clock when set to go off at a predetermined time of the day.

FIG. 3 is the electrical circuit of the entire invention, illustrated as a partial block diagram.

FIG. 4 shows schematically the attachment of the triggering mechanism lever of the clock to the switch lever of a microswitch.

FIG. 5 is a sectional view of the awakening module, showing the constructional structure thereof.

DETAILED DESCTIPTION OF THE INVENTION

Referring to the drawing shown in FIG. 1, numeral 1 designates a sound-operated switch means, having a housing 2, a sound detector or microphone 3, an electric On-Off toggle switch 4, a receptacle or phone jack 5, and an electric cord 6, which receives an electric current from a 115-volt 60-cycle household current source through the plug 7. An awakening unit or module 8 is detachably connected to the sound-operated switch means 1 through plug 9 and cord 10 to receive energizing current therefrom. An electronic circuit to amplify the electrical signals and to use the amplified current thereof to operate a relay therein is enclosed within housing 2, and will be presently discussed in detail using the schematic circuit diagram shown in FIG. 3.

When a sound signal is received by microphone 3, the sound signal is converted thereby into an electrical signal and fed into the electric circuit therein, wherein said electrical signal is amplified, electrically processed and fed to the telephone jack 5, from which the processed signal is transmitted to the awakening unit 8 through plug 9 disposed at the distal end of electric cord 10 for operation of said awakening unit 8.

In FIG. 2, the sound-operated switch means 1 is provided with the same electrical components as that shown in FIG. 1, with the exception that subsequent to the amplification of the electrical signal therein it does not undergo further electrical processing, since the section of the circuit for processing of the amplified signal is located in the clock housing 11 compartments designated by the broken lines 12 and 13; furthermore, the sound-operated switch means 1 is electrically connected to the electric clock 14 circuit through a cord 15, one end of which is permanently attached to the current output portion of a circuit within the housing 2 and the opposite end thereof is detachably connected to the electric clock 14 circuit through the plug 16, which is inserted into the mating socket (not shown in the figure) located in the left-side wall of housing 11. The electric clock 14 is energized by a 115-volt alternating current received through the cord 17 and plug 18 connected to an external source of 60-cycle current. The housing 11 is provided with a socket or telephone jack 19 in one wall thereof to accept the plug 9 of the awakening module 8, which receives an energizing current from the circuit connected to the current energizing the clock 14. The operation of the awakening system is detailed in the accompanying description of the schematic circuit diagram of the entire system displayed in FIG. 3.

In the schematic circuit diagram shown in FIG. 3, the silent awakening system receives through microphone 3 an audio signal in the form of a sound from any source, such as a smoke/fire detector alarm, a door bell, a telephone bell, or a similar sound source, and transforms the sound signal into an electrical signal of varying frequency, depending on the nature of the sound; the electrical signal passes through capacitor 20 and through leads 21 and 22 into a preamplifier AMP-1. The electrical output of preamplifier AMP-1 is fed through conductors 23 and 24 and capacitors 25 and 26 into a variable resistor 27, which is used to adjust the sensitivity of the system to incoming sound signals. The electrical signal from resistor 27 is fed through conductors 28 and 29 into a terminal amplifier AMP-2, in which the electrical signal is further amplified and transmitted into the solenoid coil 30 of a relay 31; this action closes the relay switch 32, thereby permitting a current flow through conductors 33 and 34 from the rectifier section 35 into the current-processing section 36, enclosed by broken lines.

In the current-processing section 36 of the system circuit shown in FIG. 3, the current from the rectifier 35 (which is the current source for energizing all sections of the entire awakening system) passes through conductors 37 and 38 into a double-pole double-throw (DPDT) switch 39. Depending on the switch contact position of switch 39, which is manually operated to select a current therefrom, the current flows through either switch section 40, such as when section 40 is closed, as shown in FIG. 3, or through switch 41, when closed (thereupon, switch section 40 will be open).

If switch 39 is manually actuated so that the switch section 41 is closed, the current from conductor 38 flows through switch section 41, the current interrupter 42, into one side of telephone jack designated by 5,19, which are respectively shown in FIGS. 1 and 2 and in the schematic circuit diagram of FIG. 3. When the awakening module 8 is connected to telephone jack 5,19, the current from the jack flows through conductor 43 of electric cord 10 into the awakening module 44,8, wherein after passing through motor 56 winding 45 (FIG. 5) the current emerges into the cord 10 through conductor 46 to the other side of telephone jack 5,19, and conductors 47 and 43 to the relay switch 32 through the conductor 33. From the relay switch 32 (which is closed now, because of the audio signal received by microphone 3, as stated previously) the current passes through conductor 34 to conductors 49 and 50 back to the rectifier 35, thus completing the circuit, and energizing the awakening module 44,8. The energized module 44,8 will produce an undulatory, pulsative effect externally thereto. This latter effect creates the awakening stimuli in a person sleeping on a pillow with the awakening module 44,8 thereunder.

Following the descriptive scheme given in the preceding paragraph, if the switch contact section 40 of switch 39 is closed manually (opening contact section 41), the current from conductors 37 and 38 will flow through the variable resistor 51 (also shown in FIG. 2) to one side of telephone jack 5,19, conductor 43 of cord 10 into the awakening module 44 or 8. In this case, the continuous flow of current of variable intensity, due to variable resistor 51, into the awakening module 44,8 will produce a continuous vibration of the housing 52 (FIG. 5). This vibration can be utilized to massage tired muscles for relaxing and soothing them. Thus, by adjusting the vibration intensity by means of resistor 51, a person can obtain the desired vibrational intensity from module 44,8.

For employing the awakening system by means of an electric clock, the electric clock 14 is set to the desired time of awakening; the time is set the same as by an ordinary electric alarm clock. When the time of awakening arrives, the alarm-triggering mechanism of the clock 14 (see FIG. 4), which is connected to a normally-open miniature electric switch 53, such as a commercially available microswitch, closes the microswitch 53, which is represented by the open switch 53 in the schematic circuit diagram shown in FIG. 3. Upon closing of the microswitch 53, a current flows from the rectifier section 35 through conductors 37 and 38 to contact section 41 of the DPDT switch 39, since for awakening purposes the switch 39 is manually actuated so that the contact section 41 thereof is closed. The current from the contact section 41 of switch 39 is fed into a current interrupter 42, which interrupts the flow of the current from 30 to 60 times per minute, depending on the manufacturer's current interrupter type (which may be a commercial light flasher, a bimetallic thermosensitive switch, or a multivibrator). The interrupted current flows through conductor 55 to one terminal of telephone jack 5,19, from which it flows into the awakening module 44,8, and to one terminal of the energizing coil 45 of a motor 56 (FIG. 5), and leaves it through the other terminal conductor 46 (of electric cord 10) and passes through the opposite terminal of telephone jack 5,19 to conductors 47 and 57 to microswitch 53, which is located in abutment with the alarm-triggering lever 59 of the clock 14 mechanism (FIG. 4). The current continues from microswitch 53 through conductors 54 and 50 and returns to the opposite side of the rectifying section 35, thus completing the electrical circuit, and thereby exciting the awakening module 44,8 to an undulatory pulsative motion, which produces an awakening action in a person using the awakening module.

FIG. 4 illustrates the mechanical connection of the microswitch 53 lever 58 to the alarm-triggering lever mechanism 59 of clock 14. The solid lines by which the microswitch lever 58 and alarm-triggering lever 59 are shown illustrate the closed position of the microswitch 53, the projection 60 being current-closing lever of the microswitch 53. The structures designated by 59a and 58a, shown by broken lines, illustrate the normally open (inactive) positions of the respective structures 59 and 58. Numerals 54 and 57 designate the output conductors of the microswitch 53, also shown in FIG. 3. The designations 61 and 62 limited by broken lines respectively indicate the compartments 12 and 13 within the clock 14 (see FIG. 2); these compartments accommodate the electrical parts forming the respective circuits shown in the schematic circuit diagram of FIG. 3. The microswitch 53 closes when the alarm-triggering lever 59 is actuated by the clock mechanism when the time for awakening arrives.

FIG. 5 illustrates the arrangement of the various structural components of the awakening module 44,8. The electric current enters the module through conductors 43 and 46 enclosed in the sheathing of cable 10. The conductors carry either a continuous current, when the section 40 of DPDT switch 39 is closed, and a pulsative or repeatedly interrupted current when the section 41 of DPDT switch 39 is closed. In either case, the electric current energizes the direct current (DC) motor 56, causing its armature 63 together with its shaft 64 to rotate at high speed, up to 4000 RPM. At one terminal

portion of shaft 64 is a metallic lump or weight 65 whose greater mass extends radially to one side of shaft 64 to produce an out-of-balance rotation of the shaft. This unbalance causes a vibrational effect in motor 56 during the shaft rotation. Since the motor housing 66 is embedded in a potting compound 67, such as a casting resin which is available commercially, in the housing 52 of the module 44,8, the vibratory effect of the motor is transmitted to the housing 52. This vibrational effect is used to relax and sooth tired muscles. However, when the current transmitted into the motor 56 is periodically interrupted, as by current interrupter 42, the recurrently interrupted current in motor 56 produces an additional unbalance in module 44,8; the two unbalanced forces acting at right angles upon the module housing 52 create a resultant force, which is undulatory and pulsative, similar to throbbing, causing the housing 52 to execute throbbing motion of moderate intensity. By making a slight bend in the shaft 64 adjacent the mass 65, the pulsative effect is magnified. The latter effect is utilized to awaken a person, when the module 44,8 is placed under the pillow of the sleeping person.

As stated earlier, the system receives a current from an external source, such as a 115-volt 60-cycle household current, and converts it into a nominally 6-volt direct current for the operation of all cooperative parts in the circuit. A step-down voltage transformer 68, whose primary winding 69 is connected to the 115-volt alternating current; its secondary winding 70 is connected across the two terminals 71 and 72 of the rectifier section 35, which comprises the four diode rectifier arranged in a full-wave connection. Other rectifiers, such as copper oxide plates can also be used.

It is thus seen from the preceding description that the silent (without sound) awakening in a sleeping person can be brought about by either an electric clock set to a predetermined time to trigger the awakening system or by the sound of a smoke/fire detection alarm, a door bell, ringing of a telephone bell, or other similar sounds sensed by the sound-operated switch means herewith described.

The disclosure of the invention described hereinabove represents the preferred embodiments of the invention; however, variations thereof, in the form, construction, and arrangement of the various electronic components thereof and the modified application of the invention are possible without departing from the spirit and scope of the appended claims.

I claim:

1. A silent awakening system, comprising: a first means adapted to receive an acoustic signal and to convert said acoustic signal into an electrical signal, an electric amplifying means connected to said first means and having an electric circuit therein to amplify and transmit said electrical signal to an electric relay means for operation thereof, a second means, in electrical connection with said first means, receiving an alternating current from an external source and adapted to convert said alternating current into a direct current for energizing said first means, and a third electric circuit means connected to said second means to receive therefrom a direct current through said electric relay means therein to electrically process said direct current; said third electric circuit means having in the current-processing circuit thereof a current-interrupting means, a variable current control means disposed therein in electrically parallel relation to said current-interrupting means, and a current-switching means, connected be-

7

8

tween the electrically parallel sections comprising said current-interrupting means and said variable current control means, to channel the direct current supplying said electrically parallel section into one or the other of said electrically parallel sections; and, an awakening module in electrical connection with said third electric circuit means through said current outlet means to receive an electric current therefrom for operation of said awakening module by the actuation of said electric relay means upon reception of an acoustic signal by said first means.

2. A silent awakening system as defined in claim 1, wherein said first means comprises a housing with a compartment therein, an electric circuit is disposed in said compartment and receives an electric current from an external source, an electric relay provided in said electric circuit at the output section thereof, an acoustic receptor means disposed in the wall of said housing and connected to said electric circuit for transmission of an electric current therefrom to said electric relay for actuation thereof when said acoustic receptor means receives a sound signal from a fire-detector alarm, a telephone bell, and the like.

3. A silent awakening system as described in claim 1, wherein said second means comprises a step-down transformer having the primary coil thereof connected to a 115-volt alternating current and the secondary coil thereof connected to a full-wave rectifying circuit in electrical connection with said first means to energize said first means for actuation of the electric relay means connected thereto.

4. A silent awakening system as described in claim 2, wherein said housing is provided in one wall thereof a switch means electrically connected between the electric circuit disposed in the compartment of said housing and the electric relay to activate and deactivate said electric relay.

5. A silent awakening system as described in claim 2, wherein said electric circuit disposed in the compartment of said housing is provided with a current-processing section connected in electrical relation to said electric circuit through said electric relay to receive an energizing current therefrom when an acoustic signal is received by the acoustic receptor means thereof disposed in the wall of said housing.

6. A silent awakening system as described in claim 5, wherein said current-processing section of the electric circuit disposed in said housing is provided with an electrically parallel circuit section for processing the current received thereby through the electric relay and a current outlet means in the form of a telephone jack connected to said electrically parallel circuit section is disposed in the wall of said housing whereby the processed current in said current-processing section can be transmitted externally thereto through said telephone jack.

7. A silent awakening system as described in claim 6, wherein said current outlet means receiving a processed current from the current-processing electrically-parallel circuit section transmits said processed current to an awakening unit detachably connected thereto through the telephone jack for energization of said awakening unit.

8. A silent awakening system as described in claim 1, wherein said silent awakening system is provided with means to house said first means, said second means, and said third electric circuit means electrically connected theretogether to produce an output signal therefrom,

means to receive said output signal for transmitting it to an awakening unit detachably connected to said means receiving said output signal to produce in said awakening unit an undulatory motion for awakening a sleeping person employing said awakening unit under his bed pillow.

9. A silent awakening system as described in claim 5, wherein said housing is provided in the wall thereof with a double-pole double-throw switch means connected to the electrically parallel circuit section disposed in said housing to transmit from said electrically parallel circuit section a current from either branch circuit thereof to the telephone jack disposed in the wall of said housing.

10. A silent awakening system as described in claim 1, wherein said first means is provided with a housing having therein an electrical circuit comprising an amplifying means, an electric relay connected to said amplifying means, a current-rectifying circuit receiving an alternating current from an external source to convert said alternating current into a direct current of lower voltage than that supplied by said external source and to transmit said direct current to said amplifying means for energization thereof, a current-processing circuit connected to said current-rectifying circuit and receiving a direct current therefrom, and a current outlet means disposed in the wall of said housing and electrically connected to the output section of said current-processing circuit to receive a processed current therefrom; an awakening module having an electric cord extending therefrom with means at the distal end thereof for connecting said awakening module to said current outlet means to receive an energizing current therefrom.

11. A silent awakening system described in claim 10, wherein said current-processing circuit comprises two sectional circuits connected theretogether in an electrically parallel relation; one of said sectional circuits has in the circuit thereof a current interrupter to interrupt the current passing therethrough from 30 to 60 times per minute, and the other sectional circuit is provided therein with a variable resistance means for increasing or decreasing the current intensity therethrough; said sectional circuits have a common current outlet means disposed in the wall of the housing of the first means for transmitting through said common current outlet means either an interrupted current or a continuous current of varying intensity as adjusted by said variable resistance means to an awakening unit detachably connected to said common current outlet means for energization of said awakening unit.

12. A silent awakening system as defined in claim 1, wherein said silent awakening system comprises a sonic means for detecting audible signals and converting said audible signals into electric signals, means, connected to said sonic means, for amplifying said electrical signals and having in the circuit thereof an open-circuit electric relay means, means for receiving a supply current from an external source and being in electrical relation with said sonic means and said means for amplifying said electrical signals for energization thereof, an electric timing unit having therein an alarm-triggering means and a current control means mechanically connected to said alarm-triggering means for actuation thereby; said electric timing unit is connected to said means for receiving a supply current from an external source, and an electric circuit means connected to said means for amplifying said electrical signals and adapted to receive the amplified electrical signals therefrom for processing

tnem to an electric current form usable by an awakening means in electrical connection with said electric circuit means through a detachable electric conductor thereof; said awakening means is energized when an electric current from said electric circuit means flows thereinto upon actuation of either said current control means by said alarm-triggering means disposed in said electric timing unit or of said open-circuit electric relay means receiving an electrical signal from said means for ampli-

January 16, 1978
Orange, California

fying electrical signals upon detection by said sonic means of an audible sound from an external source.

13. A silent awakening system as defined in claim 12, wherein said awakening means comprises means for producing therein a throbbing action and thereby creating awakening stimuli in said awakening means, said awakening means being adapted to be placed under the pillow of a sleeping person for transmitting thereto, through said pillow, awakening stimuli produced in said awakening means.

* * * * *

Chapter 7
Patent Office Examination
of the Application

Subsequent to the filing of the patent application in the patent office, the office personnel in the receiving department immediately date the application. The application then is classified and sent to the proper department or group for processing by the examiner specialized in the particular class of inventions. A filing receipt is prepared containing the name of the inventor, date of filing, the assigned serial number to the application, and the examining group number. A copy of this receipt is sent to the applicant for his records. It takes from four to six weeks before the receipt is mailed to the applicant.

When the applicant receives the receipt, he may mark his invention "Patent Applied For," if he is manufacturing the product. In no case should he mark the product "Patent Pending," as some manufacturers do carelessly. A patent is considered to be pending when at least a first action has been taken on the application; before then a patent is not pending. Anyone violating this rule by marking his invention "Patent Pending" and "any word importing that an application for a patent has been made, when no such application has been filed, or if filed it is not pending, to deceive the public, shall be fined not more than $500 for every such offense. Any person can sue the violating party for penalty, in which event one-half shall go to the person suing and the other half to the use of the United States." (Section 292, USC Title 35, Patents.)

The examiner assigned to the prosecution of the application makes a thorough search of the prior art as available in the Patent Office Library. The search includes the examination of the United

States patents, foreign patents, periodicals, books, and any other publications that present information pertinent to the invention and have been printed earlier than one year before the patent application has been filed. The examiner's responsibility is to understand the invention for which the application has been filed and to compare other prior inventions for determining novelty, originality, and utility of the new invention over the existing art. He also determines if the application is in compliance with legal requirements and patent office rules, as well as with formal matters. In recent years, the time needed for the first action by the examiner has been running from six months to several years. It depends on the complexity of the invention and the work load in the group to which the application has been assigned.

FIRST OFFICIAL ACTION BY THE EXAMINER

Upon completion of the examiner's study of the invention and examination of all the prior art, he notifies the applicant, in writing, respecting his findings and decision on the case. His official action contains references to the prior patents, publications related to the invention that are earlier than one year before the filing date of the application, and any reasons for rejection or objection, or any mistakes in the application. He lists his findings under a heading "Art Relied Upon" or "References of Record" and sometimes he will list matter under "Other Pertinent Art." He further points out any errors in the drawings, specification, and claims essential to the applicant's further pursuance of the case.

The objections or rejections, if any, to the claims may be based on prior art, incompleteness, indefiniteness, or the concept itself. The applicant will have the opportunity to correct or complete the requirements by amendment of the claims. The examiner's action usually will be complete concerning fundamental defects in the application. The action of the examiner may be limited to matters of form for the time being until one or more claims are found allowable. At times, the examiner will allow some of the claims and reject others or object to the language of the claim and request correction. This latter action is not a rejection, but a requirement for improvement and sometimes for the strengthening of the claim by amendment. At times, the applicant will detect his errors and request correction in his response to the examiner.

In the event the examiner finds no significant matter in the claims for objection or rejection, he allows all the claims as filed.

When references are cited, the examiner mails copies of the references together with his letter to the applicant. The applicant is given three months from the date of the office action to comply with the examiner's requirements. If the applicant fails to respond within the specified time, the application is abandoned.

In rejecting any claim, the examiner states such terms as "the structure is fully anticipated," "it is fully met by the referenced prior art," or "the invention does not present novelty." Other examples of objections might be that the structure is not patentable over the certain cited patents, it is obvious to use A's patented element designated by numeral (X) in the applicant's structure, or the invention claimed is not patentably different from prior art.

The inventor should not become disappointed and give up the case. It is possible that the wording of the claim requires slight amendment to overcome the rejection. For example, suppose the examiner states that, "a means for producing light is anticipated by A's patent because it has a fluorescent light. If the inventor's light is gas-mantle type, he can modify his claim stating, "means employing a gaseous element to produce light," to overcome the examiner's rejection. This is provided, of course, no prior art exists in mantle-type gaseous lamps.

The examiner might also object to the use of broader terms for a claim when the field of the invention is crowded with prior art. If the applicant insists that his invention has novel features not existing in prior art, he should narrow his claim by specifying the part by name so that it narrows the field to that particular part. If the claim is so narrowed for allowance of a claim, however, the claim might become useless because the invention will be restricted and the market for the product might become minimum.

The applicant would have a second chance to introduce his claims in amended form to the examiner, even after the examiner states "this action is final," if the applicant can produce reasonably good cause for why he thinks his invention avoids prior art. A claim or claims must be amended to correct the examiner's allegation.

APPLICANT'S STUDY AND RESPONSE TO OFFICE ACTION

In amending an application, the applicant must fully and clearly point out the patentable features and novelty of his invention in the face of existing prior art which the examiner finds to be anticipating the disclosure in the application. The applicant, regardless of the examiner's action, may make minor amendments in his specification within the scope of his invention, as well as in his claims. He may amend or cancel a claim or claims and substitute new claims, or cancel a word or a phrase or clause and substitute a new word, a phrase, or a clause. He should always keep in mind—in his cancellation and substitution—that new material is within the scope of the original concept in compliance with his drawings and specification.

To complete the patent office action with regard to the patent application Serial No. 869,797, now patent No. 4,180,810 (shown in Chapter 6), the examiner's first action dated November 24, 1978, stated in the summary of action of his letter that Claims 1 through 21 (original number of claims) are rejected on grounds of Fossard patent 3,786,628, Class 240, and Subclass 407, and Muncheryan patent No. 4,028, 882, Class 340, and Subclass 407. The examiner further stated in a short note:

"This application contains Claims 1 through 21. Claims 1 through 21 are rejected over Fossard taken together with Muncheryan (applicant's patent). This application appears to be an improvement over the (applicant's) patented system wherein the vibratory unit is also triggered acoustically. However, Fossard teaches the concept obvious to employ in combination with the Muncheryan system. No claim is allowed."

Signed
Examiner Harold Pitts

Such a letter from an examiner is quite common. The applicant must not be dismayed because it is the examiner's duty to point out the closest prior art to the applicant. The responsibility of proving that the applicant's invention is not anticipated by the cited prior art is that of the applicant; his argument should contain facts. After studying the remarks and references to the prior patent and to the Fossard patent, the applicant responded to the examiner's letter as follows:

AMENDMENT

IN THE UNITED STATES
PATENT AND TRADEMARK OFFICE

H. M. Muncheryan
Serial No. 869,797   Group Art Unit 234
Filed: January 16, 1978   Examiner Harold Pitts
For: Silent Awakening Device
January 2, 1979

To the Commissioner of Patents and Trademarks:

This communication is in response to the office action dated December 7, 1978. (The examiner dated his letter on November 24, 1978 but it did not leave the patent office until December 7, 1978.) The matter in parenthesis is not included in the applicant's response; it is only an explanation to the reader about the differences of 'he dates given above.

SPECIFICATION

Fossard Patent No. 3,786,628, with a single claim, reads on an old alarm clock, invented centuries ago, whose buzzer vibrates upon arrival of set time, and when the clock is placed under a pillow it awakens the sleeper by the buzzing vibration, as claimed. Any person is free to purchase such a clock (no patent exists on it) and set it to a predetermined time for awakening, and place it under his pillow to awaken him by the noise of the buzzer. No one needs a patent on such an old contraption covered by prior patents a century ago. Therefore, the applicant does not believe that the Fossard disclosure anticipates on the applicant's invention from any standpoint.

. . . Simulating the Fossard language in his patent disclosure, the applicant, by way of an example, offers the following "presumably invented" disclosure (see Fig. 7-1). Assume that the applicant has thought of an idea of substituting a fuel or energy for use in an automobile engine, because of the shortage and high cost of gasoline. The applicant states (in his application): The applicant has conceived a new and useful fuel substitute for use in automobile engines. Said substitute comprises a source of lunar energy designated by 1 in the drawing of Fig. 7-1, solar energy 2, and gravitational force energy 3, connected together as shown to an automobile engine 7 through the leads 4, 5, and 6, respectively. Numeral 8

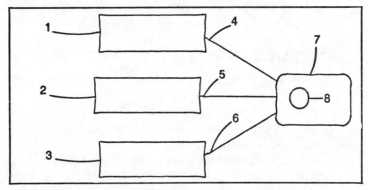

Fig. 7-1. Substituting energy for gasoline in an automobile.

designates means for "identifying energy emitted and selective control thereof" (per Fossard language).

No patent can be granted on such an incomplete, vague, indefinite, and obscure idea, since no operative means has been disclosed . . . . . Accordingly, in the absence of an operative structure, description, or any claim covering any acoustic detection and processing device, anyone has the right and privilege by patent law to think of the idea of triggering a device with sound signal and *inventing a workable system* for the detection of an acoustic signal to trigger a device, such as the applicant's invention, to an operational state. The applicant has successfully accomplished this achievement. Thus, the applicant has avoided any disclosure that Fossard has presented as his invention and that the office letter relates the same to the applicant's invention and application. Fossard has failed to give any description and details of an invention for enabling anyone skilled in the art and science to which the invention pertains to make and use the invention. Not only the Fossard invention discloses nonsense, but the idea of trying to pass this through the examiner's desk with such nonsensical imagination would be plain insanity, and, of course, no patent can be granted on such nonsense.

With regard to Muncheryan patent, No. 4,028,882, the applicant has cancelled those claims he considers to have certain antecedent relationship to the claims included in the applicant's prior patent No. 4,028,882, and has retained those that are pertinent to the improvement shown and described in detail in the present application. The applicant submits that he has placed the present application in a form for allowance.

Further and favorable consideration is respectfully requested.
Orange, California
January 2, 1979

Signed _____
Hrand M. Muncheryan, Applicant

EXAMINER'S SECOND OFFICE ACTION

On March 22, 1979, the applicant received the following letter from the examiner in charge of the case.

"All of the claims being allowable, the prosecution of the merits is closed in this application and the NOTICE OF ALLOWANCE will be sent in due course."

After the notice of allowance, no communication should be directed to the examiner, because it is not needed, and this will delay further action on the application. Applicants must remember these points.

EXAMINER'S FINAL DISPOSITION OF THE CASE

The notice of allowance was issued on May 21, 1979. A payment was made on July 28, 1979. At that time the applicant also requested, and included the fee for, 12 additional copies of the patent (when issued). The patent was issued on December 25, 1979.

HOW TO AMEND CLAIMS

There is a certain format that has been established by the Patent and Trademark Office to facilitate the amending of claims when an applicant is responding to the examiner's action. The following examples will illustrate this.

Let's assume that the following two claims contained words that had to be altered because either the examiner objected to their presence or the applicant noticed them while complying with the examiner's action. The word *diodes* in Claim 4 and the words *connected* and *intended to hold* contained in Claim 5 are to be amended. The rule for deleting a word is as follows. The word to be deleted is placed between parentheses, and a word or phrase to be inserted is underlined and included between dashes at each end. For the benefit of the reader, the words or the phrases to be changed in the Claims 4 and 5 are included in italics as follows.

Claim 4. A security surveillance laser system as defined in Claim 1, wherein the *diodes* having a second means are devices capable of converting the incident photonic energy thereupon from the split laser beams into an electrical energy to perform mechanical operation in said second means.

Claim 5. A security surveillance laser system as defined in Claim 1, wherein said second means electrically *connected* to said first means and *intended to hold* a quiescent operational state in said security surveillance laser system is a combination of electrical relays and switches responsive to the electrical energy resultant from said first means.

AMENDMENT OF CLAIMS

Claim 4, Line 2: Cancel the word (diodes) and substitute therefor—radiation-sensing elements—.

Claim 5, Line 2: Cancel the word (connected) and substitute therefor—adapted to sustain—.

Line 3: Cancel the phrase (intended to hold) and substitute therefor—adapted to sustain—

PTO 76-13a
(Rev. 3-77)
(Formerly PTOL-85)

All communications regarding this application should give the serial number,
date of filing, and name of the applicant

U.S. DEPARTMENT OF COMMERCE
Patent and Trademark Office

Address: COMMISSIONER OF PATENTS AND TRADEMARKS
Washington, D.C. 20231

NOTICE OF ALLOWANCE
AND BASE ISSUE FEE DUE

The application identified below has been examined and found allowable for issuance of Letters Patent.

FILING DATE	SERIAL NO.	NO. OF CLAIMS ALLOWED	EXAMINER AND GROUP ART UNIT	
01/16/78	869797	13	PITTS	234
APPLICANT(S)	MUNCHER, HRAND M.; ORANGE, CALIF.			
TITLE OF INVENTION (X indicates as amended by examiner)	SILENT AWAKENING SYSTEM		MAILED MAY 21 1979	

	BASE FEE COMPUTATION		BASE FEE DUE	CLASS-SUB
$100.00	+ (FOR DWG @ $2 PER SHEET) $2.00	+ $10 (FOR FIRST PAGE PRINTED SPEC.)	$112.00	340/407.000

The complete Issue Fee is one hundred dollars ($100) plus two dollars ($2) for each sheet of drawing, plus ten dollars ($10) for each printed page of specification (including claims) or portion thereof.

Inasmuch as the final number of printed pages cannot be determined in advance of printing, an initial BASE ISSUE FEE (consisting of the fee for printing the first page of specification ($10) plus the fee of ($2) for each sheet of drawing, added to the fee of $100) *must be paid within three months from the date of this notice,* or the application shall be regarded as ABANDONED.

When remitting said Base Issue Fee, enclosed Form PTOL-85b should be used, and if use of a Deposit Account is being authorized, PTOL-85c should also be forwarded.

The Base Issue Fee will not be accepted from anyone other than the applicant, his assignee, attorney, or a party in interest as shown by the records of the Patent and Trademark Office.

If an assignment has not been previously filed and it is desired to have the patent issue to the assignee, the assignment must be received in this Office with the recording fee together with the Base Issue Fee. In any event, the appropriate space(s) under "Assignment Data" on PTOL-85b must be completed. Where there is an assignment, the assignee's address must be given to ensure its inclusion in the printed patent.

In connection with the address of the inventor(s), attention is directed to Form PTOL-231 enclosed.

A Notice of Balance of Issue Fee Due will be mailed together with the patentee's copy of the patent *if an additional fee is due.* Payment must be made within three months from the date shown on said Notice since FAILURE TO PAY THIS BALANCE WITHIN THE TIME SPECIFIED WILL RESULT IN LAPSE OF THE PATENT.

IMPORTANT

ATTENTION IS DIRECTED TO RULE 334
REVISED NOVEMBER 4, 1969.

THE PATENT WILL ISSUE TO APPLICANT
UNLESS AN ASSIGNEE IS SHOWN IN
ITEM 2 ON FORM PTOL-85b, ATTACHED

HRAND M. MUNCHERYAN
1735 N. MORNINGSIDE ST.
ORANGE, CALIF. 92667

YOUR COPY-See reverse side for Base Issue Fee Record

P03 BATCH-H01

Fig. 7-2. Examples of fee forms.

PTO 76-13c (Duplicate of PTO 76-13b for forwarding to Patent Office if use of Deposit Account is authorized.)
(Rev. 3-77)
(Formerly PTOL-85)

BASE ISSUE FEE TRANSMITTAL

This form is provided in lieu of a formal transmittal and should be used for transmitting the Base Issue Fee. Items numbered 1 through 4 below should be completed as appropriate. The Base Issue Fee Receipt will be mailed to the address appearing in item 4 or as designed in item 4a below.

1A. The COMMISSIONER OF PATENTS AND TRADEMARKS is requested to apply the Base Issue Fee to the application identified below and deliver the patent as indicated.	1B. For printing on the patent front page. List below the names of not more than 3 registered patent attorneys or agents OR, alternatively, the name of a firm having as a member a registered attorney or agent. If no name is listed below, no name will be printed.
(Signature of party in interest of record) (Date)	1
	2
	3

NOTE: The Base Issue Fee will not be accepted from anyone other than the applicant, his assignee, or attorney, or a party in interest as shown by the records of the Patent and Trademark Office, nor will this fee be accepted in the application prior to the Notice of Allowance.

FILING DATE	SERIAL NO.	NO. OF CLAIMS ALLOWED	EXAMINER AND GROUP ART UNIT
01/16/78	869797	13	PITTS 234

APPLICANT(S) MUICHEK, HKAND M.; ORANGE, CALIF.

TITLE OF INVENTION (X indicates as amended by examiner) SILENT AWAKENING SYSTEM

				NOTICE OF ALLOWANCE DATE
				MAILED MAY 21 1979

BASE FEE COMPUTATION				CLASS-SUB	BASE FEE DUE
$100.00 +	(FOR DWG @ $2 PER SHEET) $2.00	+ $10	(FOR FIRST PAGE PRINTED SPEC.)	340/407.000	$112.00

2. ASSIGNMENT DATA (print or type)

	BASE FEE ENCLOSED: ☐ YES ☐ NO

A. The appropriate box(es) in this item MUST be checked:

(1)		This application is **NOT** assigned;
(2)		This application **IS** assigned;
(3)		Assignment herewith;
(4)		Assignment recorded and returned by Patent and Trademark Office:

| | YES | | NO |

B. For printing on the patent: (Unless an assignee is identified below, the patent will issue to the applicant above-named. Completion of this item, however, is NOT a substitute for filing the assignment as required in Rule 334)

(1) NAME OF ASSIGNEE:

(2) ADDRESS: (City & State or Country)

(3) STATE OF INCORPORATION, IF ASSIGNEE IS A CORPORATION.

Charge to my Deposit
Account Number: _____

(PTOL-85c must be enclosed)

a. | | For Base Fee — $112.00

Patent Copies— 6.00
_____
$118.00

b. | | For Balance of Issue Fee Due, if any.

c. | | For Recording Enclosed Assignment.

DO NOT USE THIS SPACE.

MAILING INSTRUCTIONS

4a. Further correspondence is to be mailed to the following:

NOTE: All further correspondence, the patent together with the Notice of Balance of Issue Fee Due, if any, will be mailed to the addressee entered in the stub marked 4 at the lower left below, unless you direct otherwise by specifying the appropriate name and address in item 4a below right.

```
HRAND M. MUNCHERYAN
1735 N. MORNINGSIDE ST.
ORANGE, CALIF. 92667
```

TRANSMIT THIS FORM WITH PTOL-85b WHEN USING DEPOSIT ACCOUNT BATCH-1101---

Fig. 7-2. Examples of fee forms. (Continued from page 59.)

61

A sentence stating the following would be written to the examiner. "The applicant believes that the substitution of the appropriate words in Claims 4 and 5 will improve the claims by making them more definite, and clarifies the sense of each claim thereof."

ALLOWANCE AND ISSUE OF PATENT

Subsequent to the allowance of the application, a notice of allowance is sent to the applicant. The final fee becomes due within three months from the date of the notice. The basic fee (except for design cases) is $100, plus $10 for each printed page of the specification, and $2 for each sheet of drawing. Together with this fee, the applicant may include an additional fee for purchasing copies of the issued patent at a cost of 50 cents per copy. For instance, if the applicant wants to purchase 10 copies of the original patent, he includes $5 in addition to the other fees. Let's assume that six pages of the specification were to be printed and there were two sheets of drawings. The total fee to be sent to the patent office would be:

Total Fee = $100 + 6 × $10 + 2 × $2.00 + 10 × $0.50 = $169.00

A total fee of $169 should be attached to the Base Issue Fee Transmittal form, furnished by the patent office, together with the Notice of Allowance to the applicant.

The applicant signs the Base Issue Fee Transmittal form (see Fig. 7-2), attaches the fee to the completed form, and mails it to: The Commissioner of Patents and Trademarks, U.S. Department of Commerce, Patent and Trademark Office, Washington, DC 20231.

A provision is made in the statute that, if the applicant does not have the funds sufficient to pay the total fee, a request may be made to the commissioner for a delay in payment of the final fee for reasons the applicant must explain. The commissioner, upon finding the reasons tenable, may extend the time for an additional three months. When the issue fee is paid, the patent is issued as soon as possible thereafter (depending upon the volume of work at the patent office). The official patent is mailed on the day it is granted. The record of patent is then open to the public, and the printed copies of the specification and the drawing are available on that same day to any person who wants to purchase them.

GENERAL DISCUSSIONS

The preceding discussions indicate that rejection of any claim by the examiner in the first action should not be construed as final. For instance, the examiner may not view the content of your application in the same light as you do. In such a case, carefully explain to him the facts; do not become perturbed about his statements. The examiner will be willing to help you. Here are a few suggestions.

☐ Respond to every objection, suggestion, or rejection of the examiner.

☐ Never accuse the examiner of being wrong. State to him the facts the best you know.

☐ Correspondence with the examiner should be carried out with decorum.

☐ Write your response or amendment in third person.

☐ Never argue with the examiner by finding fault in his statements. Put your statements to him in the form of suggestions.

☐ In convincing the examiner about your invention being different from those in prior art, state to him that your claims answer all reasonable requirements concerning prior art.

☐ When you don't understand the examiner's statements, write to him and request clarification. Do this as early as possible during the three months of response period so that you will receive a reply within that time.

☐ End your final statement with the words "respectfully requested."

☐ Sign your full name under the, Respectfully Yours.

Chapter 8
Division of the
Parent Application

If two or more distinct inventions independent of each other are filed in an application, the examiner may require the application to be restricted to one of the inventions. In such an event, a *divisional application* must be filed. A divisional application is a continuation of the parent application. If filed prior to the issuance or abandonment of the parent application, it will be entitled to the benefit of the filing date of the parent or original application. A divisional application may result from several causes uncovered by the examiner during the prosecution of the original application. A patent issued on a divisional application may not be anticipated by the parent application or its patent if the divisional application is filed before the patent on the original application is issued.

CANCELLING PARTS OF A PARENT APPLICATION

When more than one distinct and independent invention is present in an application, the applicant will be required to select one of the inventions and to direct his description and claims to that one invention (if an allowable claim exists in the application). The other invention is filed as a *continuation application*. If no allowable claim is found in the application, then the application is abandoned or cancelled. A new application, with a new specification and claims on the invention, which could be improved as a result of findings of prior art in the patent office, may be filed with benefit to the applicant.

In the event the prior art anticipates one or more of the distinct inventions in the original application, the applicant should cancel them and pursue only the single invention that is not restricted by any of the references cited. The validity of the patent on the parent application shall not be questioned because it contained matter restricting it to one invention.

At times, it might be advantageous for the inventor to include more than five species in his parent application to benefit from the filing date of the first application. When the prosecution of the case is advanced in the patent office, then the inventor may file a divisional application on those in excess of five species. The divisional application will bear the original file date.

In still another case, the examiner might declare that two distinct inventions exist in one application and requires the applicant to restrict his application to one of the inventions. In such a case, if no allowable claim is found on the elected invention, then the applicant may cancel the elected invention and apply for the other invention in a divisional application, which will bear the original filing date, if the application is filed before the elected invention is cancelled.

Upon imposition of a requirement for restriction to five species in an application, the applicant must cancel all the claims reading on the nonelected species and pursue the claims that cover the first five elected species. The applicant must file a divisional application prior to the issuance of a patent on the five species in order to conserve the original filing date in the continuation application. If after filing the divisional application the claims in the continuing application are allowable, they cannot be rejected on the basis of allowed claims on the five species contained in the parent application. In the event, the claims in the divisional application are not patentable because of anticipating prior art, the examiner will not require the applicant to file a divisional application (thus saving him the extra cost of filing).

DRAWINGS IN A PARENT APPLICATION

In a divisional application, if the drawings of the new application will be identical with those originally filed, then the divisional application may include the same drawings if no conflict in the description of the two applications exists. If the prior application is or is about to be abandoned, the applicant may request the transfer of the drawings to the new application. When those drawings that are to be used in the parent application are cancelled, then the remaining drawings can be used for the new application. The cancelled drawings may remain on the sheets of the parent application, but no reference to them is made in the new application.

ADDITION OF NEW MATERIAL TO A DIVISIONAL APPLICATION

When new material is to be added to the divisional application, the new application cannot retain the original (parent) application date. In such an event, it will be best to make new drawings of the new and improved invention. New claims must be drafted to cover the new species in the application as if no prior patent application by the inventor had been filed. If the applicant's prior patent application is still pending or has become a patent, the examiner cannot cite the prior patent or the application against the second application if the statutory period of one year has not expired.

FILING OF A DIVISIONAL APPLICATION

A divisional application must contain the same parts as the parent application. It must contain a petition, abstract, specification, claims, and oath. A drawing should accompany these component application parts unless the original drawing is to be used with the divisional application. In addition, the filing fee—as required by the statutes—should accompany the application at the time of filing. Otherwise, the handling and classification of the application might be delayed in the patent office. The preamble of the divisional application should contain a clause to identify the relationship between the divisional application and the parent application. This can be achieved by stating that the application of Serial No. XYZ (parent application) has been divided and the division of application has formed the present continuation application. In this way, the two applications are linked.

REVIVAL OF AN ABANDONED APPLICATION

An application that is abandoned for failure to prosecute or for other reasons may be revived as a pending application if the application is accompanied by a statement showing, to the satisfaction of the commissioner, that the delay was unavoidable. A petition to revive the abandoned application must include acceptable reasons such as the applicant was absent from home for business or personal reasons due to applicant's illness or for some other unavoidable reason. The delay might also have been due to failure of the post office to transmit the application to the patent office on time. In such a case, the mailing date must be shown to be prior to the expiration date of the satutory period set by the last letter from the examiner. The petition to revive an abandoned application should be accom-

PETITION TO REVIVE AN ABANDONED PATENT APPLICATION

IN THE UNITED STATES PATENT AND TRADEMARK OFFICE

John J. Smith)	Patent Group 234
Serial No. 333,555)	Date________
Filed: March 5, 1978)	
For: A Silent Awakening)	
System)	

To the Commissioner of Patents and Trademarks:

In response to the Office Letter dated July 24, 1979, wherein the Examiner holds the applicant's above-identified patent application as abandoned because of the expiration of the statutory period of three months, the applicant submits that the delay was unavoidable for the following reasons:

(Give a reasonable cause of delay here.)

City of________________	Signed ____________________
State of________________	Applicant's full signature.

Fig. 8-1. An example of a petition to reverse an abandoned patent application.

panied by a fee of $15 payable to the Commissioner of Patents and Trademarks. An example of such a petition is shown in Fig. 8-1.

Verified causes of delay are usually sufficient to revive an abandoned application for a patent. Any inaccurate or falsified statement made in the petition may constitute a cause for rejection of the petition or nullification of the patent when issued. When the commissioner has denied the petition, then a new application containing the matter found in the parent application should be filed with a $65 fee.

TYPES OF ABANDONMENT

There are several types of abandonment of inventions. The inventor must especially guard against legal abandonment of his invention. If this happens, the abandoned invention becomes public property and the inventor cannot recover the status of patentability of the invention.

Intentional Abandonment

After an inventor constructs a working model, he might decide not to pursue its exploitation for commercial use. If the inventor is a humanitarian person, he might dedicate his invention to public use because he does not want the invention to be wasted by discarding it. He may want to take out a patent on it and assign the patent to the United States Government or Free Use of People (or both). In such an event, the inventor feels he is contributing to the welfare of his country and does not mind spending a few hundred dollars for the patent. The patent is issued to the name of the government or in favor of general public as assignee. This type of abandonment is irreversible and cannot be revived if the inventor later decides to work the invention by excluding others. He can make and sell the invention, but he will not have any monopoly over it.

Abandonment Due to Neglect

When an inventor has constructed a working model of his invention and has put it aside and has done nothing about it, he will be barred from getting a patent as a result of the one-year rule. He is considered to have exercised negligence in not applying for a patent after perfection of his invention. He is also said to have kept secret his invention by delaying its commercial use or failing to apply for a patent.

If the inventor has filed an application for patent in a foreign country and has been granted a patent before a United States patent application has been filed and issued, then the 12-month statutory period will have expired and the inventor cannot obtain a patent in the United States. His invention is abandoned by reason of neglect. If the United States patent is issued before the foreign patent, then no abandonment occurs. An inventor must be diligent from the time of conception of his invention until a patent has been issued to him.

Statutory Abandonment

When an inventor contemplates bringing a legal suit against an infringer, he should first study his patent carefully to determine if one or more claims have been anticipated by a prior patent or patents. If such is the case with respect to any claim, he must first file a disclaimer to cancel the claim or claims in his issued patent that contains the defect. This action is a simple notification by a letter to the commissioner asking him to cancel those claims the inventor thinks do not rightfully represent his invention. If he fails to file a disclaimer and secure a cancellation of these claims, then when he has initiated a court action and one or more of his patent claims are found to have been anticipated by prior art, he will lose the case. The result could be that his patent would be declared invalid.

To summarize, the causes of abandonment of a patent application can arise because of the failure of the inventor to respond to the examiner's action within the time alloted or he might fail to pay the final fee on time. An inventor might not pursue the application because he might not consider his invention as important as he first thought and not worth the cost (final fee). Therefore, he voluntarily abandons the application.

Nevertheless, if the inventor finds himself capable of meeting the financial costs, and has possibly improved on the invention, he may file a new patent application regardless of the abandoned application status. His application then will cover his new product. If any question of abandonment arises, the inventor can prove that he was diligently pursuing the improvement of a better and improved invention, and one that would avoid anticipation of prior art, now that he has become familiar with the existing art in the field of his patent.

Chapter 9
Interference
of Patents

An *interference of invention* is instituted when two or more pending applications in the Patent and Trademark Office conflict and the question of priority is to be determined. In an interference case, at least two parties claim substantially the same patentable invention. The patent office determines which party is the first inventor by examining the evidence presented by both inventors. An interference proceeding may also be initiated between a pending application and an issued patent. The most recent patent office reports indicate that about 1 percent of all the patent applications filed become involved in an inteference proceeding.

The interference proceeding is instituted after allowable claims have been found in the interfering applications that define the same patentable subject matter. No interference can be declared between the applications of the same party or between the pending application and the issued patent owned by the same party. When an interference is declared, the parties involved in the proceeding must declare the ownership of the inventions. Any changes made in the right, title, or interest affecting the ownership subsequent to the declaration of interference must also be made known to the patent office.

The examiner forwards copies of the allowable claims to the rival applicants. The applicants are required to furnish the following information: the date of conception of the invention, date of the first drawing, date of the first description written, the earliest date of disclosure to others, date of construction of the first working model, and the date when the inventor began to improve his invention. In case the inventor has filed a foreign application, the date of foreign filing is also included in the applicant's report. Accordingly, one can readily see why it is vital for the inventor to keep an accurate record of all his developments from the date of his conception of the idea until an application is filed. In the United States, it is the person who has invented (put the idea into operating condition) the product first that receives a patent—provided all other statutory conditions have been met.

PURSUING INTERFERENCE PROCEEDINGS

Both parties in interference must gather their evidence from the time of the conception of the invention, through its development, to the time when a patent application was filed. The party who has filed his patent application first in the patent office is considered to be the senior party. The one filing his patent application after the first party is known as the junior party. The burden of proof of priority rests upon the junior party.

During the interference proceeding, the question of reduction to practice arises. When the inventor has developed his invention and has made a workable model, an actual reduction to practice has occurred. Upon filing a patent application, a constructive reduction to practice occurs. The inventor must further show diligence by working on his invention to improve its certain features in the interim between the construction of the working model and the filing of the application. Negligence to continue doing nothing on the invention after a model is made may become a bar in the interference proceedings. Therefore, the inventor must show, by actual evidence in hand, that he was diligent and was improving his invention prior to filing the application. All this information must be included in the inventor's file with the interference case.

If no evidence is forthcoming, with respect to the dates of conception and model construction (by either party), the party unable to submit evidence will be restricted to the date of his filing of a patent application. Three interference examiners study the evidence submitted by each party to determine priority. The losing party may appeal to the Board of Customs and Patent Appeals or file a civil action against the winning party. The last procedure is more complicated and costly. It might be easier for the losing party to reinvent his invention, apply for a new patent, and benefit from what he has learned from other party's invention and prior art.

DETERMINATION OF FIRST INVENTOR

The commissioner will issue a patent to the applicant who is adjudged the prior inventor. It must be reiterated here that the burden of proof is upon the junior party. The inventor who filed his application first is considered to have reduced his invention to constructive practice first. The following examples of interference cases illustrate the manner of adjudgment of priority.

☐ Johnson and Thompson conceived the invention at the same time, but Johnson reduced the invention to practice first. Johnson wins.

☐ Johnson conceived the invention after Thompson, reduced to practice before Thompson, and filed for a patent application before Thompson. Johnson wins.

☐ Johnson conceived his idea in March and reduced it to practice in December of the same year, using continuous diligence by reworking his idea on paper. Thompson conceived the idea in May and reduced it to practice in September. Johnson wins. Had Johnson not used diligence, then Thompson would have won because Johnson would have been considered to have abandoned his invention in the interim.

☐ Thompson conceived his idea first and reduced it to practice first, but did nothing further. Johnson conceived his idea and filed a patent application (constructive reduction to practice) and received a patent. Thompson filed an application after Johnson. Thompson loses the case because he concealed his invention until Johnson received his patent.

☐ Both Johnson and Thompson conceived the idea and reduced it to practice at the same time. Thompson applied for application first. Thompson wins because no two identical patents can be issued on the same invention.

☐ Suppose Johnson is the senior party and failed to file a statement for the interference proceeding. Johnson's date of conception is considered to be the date of execution of his oath accompanying his application. The patent is granted to Johnson. No interference is declared to prove priority

APPEAL TO COURT OF CUSTOMS AND PATENT APPEALS

Section 141, USC 35, Patents states, "An applicant dissatisfied with the decision of the Board of Appeals (commissioner, the deputy commissioner, the assistant commissioners, and the examiner-in-chief) may appeal to the United States Court of Customs and Patent Appeals, thereby waiving his right to proceed under Section 145 of this title, Civil Action to Obtain Patent." When such an action is taken, the applicant must file his reason for appeal with the Patent and Trademark Office. The time and place of hearing, as designated by the United States Court of Customs and Patent Appeals, will be given to the commissioner and to the parties in the interference proceeding. The commissioner will also transmit in writing the grounds for the decision of the Patent and Trademark Office. The appeals court will hear and determine the appeal on any evidence previously introduced at the patent office. The appeals court then returns its decision to the commissioner, and the commissioner enters the proceedings in the case and acts thereby.

CIVIL ACTION IN INTERFERENCE CASES

When the applicant is still not satisfied with the action of the Board of Appeals, he can take civil action against the commissioner in the United States District Court for the District of Columbia. If the court adjudges that the applicant is entitled to receive a patent, it may authorize the commissioner to issue a patent on the applicant's invention—provided other requirements of the law have been fulfilled.

It must be emphasized here that, unless the case involves a basic invention that has very high commercial possibilities, the inventor will save time and money by abandoning any interference proceedings. In the event the invention has very high merits, the inventor should resort to the services of a competent patent lawyer specialized in defending interference cases. Such a person's services will be costly, but the benefits of the invention, when commercialized, will more than compensate for the financial outlay for the legal battle.

Upon determination of the winning party, the case is referred to the examiner who notifies the losing party and resumes his action with the winning party's application. Sometimes both parties gain from the result of the appeal. One party is granted some of the claims that are different from the claims of the other party and the other party has some priorities to other claims. The applicants in the interference proceedings bear all the costs incurred.

DETRIMENTS OF INTERFERENCE PRACTICE

Sometimes a company, that keeps close watch over developments on certain types of products in which the company specializes, files a patent application for the purpose of forcing the patent office to disclose the pending patents by interference proceedings. The patent might be of no value to the company, but by the continuation of the interference case they can keep the patents pending for years. Such a procedure is known as *dragnet application.*

Another aspect of an interference case is called an *intimidating interference.* An intimidating interference can result when an inventor, after applying for a patent, discloses his invention to another company for the sole purpose of selling his patent to that company or to have his product manufactured on a royalty basis. The other company that is in the business of producing similar products wants to restrain the inventor from making and selling the product. It is not interested in making a business arrangement with the inventor. The other company files a patent to force the declaration of an interference.

The patent office, as usual, has to disclose all pending patents on similar products to all parties concerned. If the second party (company) can intimidate the first party (the inventor) because of the large amount of finances involved in pursuing the interference case, and thus force the inventor to abandon the interference proceedings, then the second party reaps the fruit of the outcome. Such a case can be defeated if the inventor has kept adequate records of his invention from the time of its conception to the filing of the patent application. The inventor must use his discretion as to whether he should carry on with the proceedings (depending on the importance of his invention). A good rule to follow is not to disclose an invention to anyone before a patent is issued. It may take a year or two before the patent is issued, but the inventor will emerge a happier person if he does not become involved with a second party about whose business character the inventor knows nothing.

Chapter 10
Patent
Infringement

Any practice of making, using, or selling a patented invention during the term of the patent, without the consent of the patentee, is an act of *infringement.* Anyone who actively induces infringement of a patented invention during its useful life is an infringer. The invention can be infringed upon only in the country where it is patented during its life of 17 years. A patented article, for instance, in the United States, cannot be manufactured abroad and brought to the United States for sale except by the patent owner. On the other hand, if only the process of making the article is patented in the United States, it would not be an infringement to have the article manufactured in a foreign country and import it to the United States for distribution and sale. Also, it would not be an act of infringement if the article is produced by using a process other than that patented in the United States. A patent issued in the United States is effective only in the United States and its territories.

INFRINGEMENT PERIOD

No infringement is committed before a patent is issued or after the patent has expired. To make, sell or use an invention while a patent on it is still pending does not constitute infringement. An article marked *Patent Applied For* or *Patent Pending* cannot exclude others from practicing the invention until a patent is granted. The reasons for this are that there is no assurance that a patient will be granted on a pending patent application, and the inventor's patent protection begins from the day of granting and publication of the patent in the *Official Gazette* of the patent office.

Nevertheless, anyone who, without the knowledge of a patent on a given product, practices a patented article is committing infringement whether he is aware of a patent or not. The marking of *Patent Applied For* or *Patent Pending* is valuable to the extent that the inventor is giving notice to the public that, in the event the patent application becomes a patent, the maker, user, or seller of the patented product, without his consent, will become subject to infringement suit.

AVOIDANCE OF INFRINGEMENT

A patented article is infringed upon when one or more of the claims contained in the patent is infringed upon. A patent dominates an invention within the scope of the claims it contains. Each of the claims can be considered to be a separate patent and dominates the structure containing the parts, elements, or steps it covers in the invention. A claim does not dominate any structure that does not contain an element stated in the claim. Any infringement on a single claim that states the constituents of the invention is considered to infringe upon the patent. Practicing a part of a claim does not constitute infringement. Unless the claim has specific limitations that restrict the scope of the invention, combining the elements recited in a dominating claim with other parts to produce another article does not avoid infringement.

To expand upon the above discussion, suppose a clock manufacturer sells his clocks to a second party who adds other parts to the clock to make a time-card registering system. A third party might have a patent on a time-card registering system. If the clock manufacturer was not aware that the purchaser of his clock was adding the missing elements to make a time-card registering device, the clock manufacturer is not contributing to the infringement. If it can be proven that he was aware of the infringement at all times while selling his product to the second party, then he will be guilty of inducing direct infringement on the time-card registering system patented by the third party. The accepted and safe way of avoiding infringement on a patented article is to avoid practicing the patented invention. Otherwise you will be taking the risk of an infringement suit.

REMEDY FOR INFRINGEMENT

The only remedy for the patentee, with respect to an infringement on his patented product, is to bring a law suit in a federal court against the infringer. The court nearest the location of the infringer should be chosen. When the infringed article is being nationally distributed by the infringer, a court near the patentee's establishment can be chosen. Wholesalers, retailers, and users of the patented product are liable in the infringement suit. Because the innocent users of the patented article are liable in the suit, it would be more troublesome to the infringer if a suit is directed to several users of the patented device in widely separated districts than to bring one suit against the plant where the infringed article is being manufactured. The infringing manufacturer must defend the suit brought against the purchasers of the article.

The first step to be taken by the patent owner is not to rush to file a suit, but to send a registered letter to the infringer asking him to stop manufacturing the patented product or an infringement suit will be instituted against him. The infringer should be furnished a copy of the patent with the claims that are being infringed. The infringer should consult with his patent counsel and, if in the opinion of the counsel the patent is being infringed upon, the infringer should be advised by his counsel to stop manufacturing the article or to make a financial arrangement with the patent owner for manufacturing the patented article under a limited contract.

Some imprudent patent counsels, knowing that the patent is being infringed upon by his client, will advise him to continue the manufacture of the patented article on the basis that when a suit is brought by the plaintiff the counsel will try to prove invalidity of the patent on some unknown grounds. Remember that, regardless of the outcome of the infringement suit, the patent counsel has to be paid for his services. Therefore, it is to the counsel's advantage for the two parties to go to court regardless who wins the case. For this reason, in selecting a patent counsel, extreme care should be exercised to find an honest, trustworthy, and competent patent counsel. There are many available even though their fees are higher. In the long run, the infringer will benefit from honest and straightforward advice. Naturally, the patentee will also benefit from the honesty and integrity of the defendant's patent counsel.

Infringement suits are very costly, running from $10,000 to hundreds of thousand dollars, depending on the character of the infringed product and the finances of the infringer or the plaintiff. It is always best to first negotiate with the infringer to determine another way of concluding the case without a legal suit. Very often, when a product is important to both parties, some legally suitable arrangement can be made so that both parties benefit from the manufacture and sale of the patented product. The latter method should be diligently tried first before an infringement suit is considered. When the invention has high commercial merits, the plaintiff's counsel may undertake the infringement suit on a contingency basis and acquire an interest in the invention and its financial returns.

INJUNCTION AGAINST AN INFRINGER

When the proceedings for an infringement have been started, it will be to the benefit of the plaintiff to ask an injunction against the infringer. An injunction prohibits the infringer from making and selling the patented invention while the infringement proceedings are in progress. A preliminary injunction cannot be granted until the validity of the patent has been established. This is done only after the court decision. Furthermore, to pursue the proceedings, the patent owner must first determine whether the infringing party has sufficient funds to pursue the case, and whether the suit will warrant a judgment that will cover the cost of the suit.

If the patent owner's intent is only to stop the infringer from manufacturing his patented product, again he must be in a position to pay his attorney's fees and other related costs. If in the opinion of the patent owner and his counsel an infringement suit is not justified because of the finances involved without a possibility of recovering the costs from the infringer, it would be best not to proceed further with the case. The main consideration here is how much financial damage can the infringer cause to the patent owner in case the infringer continues to manufacture and sell the product without the patent owner's consent. This is up to the patent owner to decide.

LEGAL DEFENSE ADVANCED BY AN INFRINGER

An infringer in a patent suit may advance a number of defenses. Some of the usual defenses include:

☐ The patent is not infringed because it does not dominate the invention which the infringer is practicing.

☐ The person receiving the patent grant is not the sole inventor because he did not invent the article alone.

☐ The suit is barred by the statutes of limitations. It was not filed within the statutory six years after the first infringement was committed.

☐ The patent is invalid because of prior art (as evidenced).

☐ The inventor abandoned his invention during the life of the patent.

☐ The invention is public property because it was described in a publication one year prior to the patent application.

☐ The patent does not meet all the requirements of the statutes.

☐ The defendant was not properly notified of the infringement because he did not mark his articles of manufacture "patented."

☐ One or more claims are invalid and no disclaimer has been filed before the suit was instituted.

All these and possibly other defenses may be advanced by the defendant's attorney. They must be successfully cleared up before a decision on the case can be established.

INFRINGEMENT DAMAGES

The nature of damage to the patent owner and the amount of penalty to be paid to him can only be determined when all or most of the defenses have been carefully examined. As a result of the examination, if damages are to be awarded, a complete accounting must be made by the infringing party to the court. The liability of the defendant will be dependent on any definite losses suffered by the patent owner as a result of the competitive manufacture and sale of the invention. The amount of business conducted by the infringer on the patented product is presumed to have been taken away from the patent owner's business.

Regardless whether or not the infringer competes with the patent owner, he is liable for the profits derived from the commercialization of the patented product, as well as being liable to any financial damage to the patent owner. The extent of lowering the price of the product by the infringer and other relevant conditions enter into the calculation of the damage brought on the patent owner. Futhermore, if the infringer was notified formally of the existing patent and the infringer continued to commercialize the product, the judge may impose as much as three times the value of the actual damage.

If the infringement suit is filed after six years of statutory period, no damages can be collected by the patent owner. If the patent owner has not worked his patent during the six-year period, the infringer can be enjoined if the patent owner has marked his manufactured devices "Patented." This is a notice sufficient in itself that any infringer would be liable to damages. Unless the invention is very important commercially and the income from it would be considerable, an appeal to the Circuit Court of Appeals would not be recommended for an average inventor with average funds because of the susceptibility of losing the case.

Infringement damages cannot be reckoned by any particular formula. The amount depends on how much a patent owner can collect by showing a just cause to do so. The actual amount is also dependent on the size of the market. For instance, the first patent application on a system producing stimulated emission of radiation (laser) filed in 1960 was granted a patent after 20 years. During this interim, the laser industry has developed into a billion-dollar business. The inventor, after having spent several hundred-thousand dollars for instituting a law suit, has finally begun to collect royalties from smaller companies that cannot pursue the infringement suit because of lack of finances for this purpose. The inventor, having won the first round with the small company, begins to exploit his invention on the financially more stabilized companies. This particular case is not common, but is largely due to our present patent system and the conduct thereof. In 20 years, many improvements that are not covered by the old application have been made and the efficiency of production of stimulated emission of radiation has been increased. This overrides the earlier patent application, but apparently not in the eyes of the present patent system.

PATENT INSURANCE

Before a manufacturer purchases a patent from a patent owner, he might ask the inventor to guarantee against the liability of infringement or he might conduct his own search to establish the validity of the patent. If he is satisfied, he might purchase insurance, usually from Lloyds's of London, to protect all or part of the expenses incurred in bringing a suit against an infringer. When the search uncovers no bar against the manufacture of the patented product, the purchaser goes ahead and buys the invention. In this manner, the manufacturer minimizes his risk.

Chapter 11
Design Patents

In the preceding chapter, all the patents discussed are utility patents consisting of articles of manufacture, machines, processes, or compositions of matter. A *design patent* refers to the artistic appearance of the invention without regard to its utility. A design patent can be granted to a person having invented a new, original, and ornamental design for an article of manufacture. Such an invention emphasizes the attractiveness of appearance rather than an engineering design for an article. The statute was enacted on design patents to encourage the decorative arts (something that would give the article beauty of appearance). An article that has an utilitarian function cannot be patented under a design patent status.

The design-patent statute was established in 1842 to provide protection to the citizens of this country for their efforts in producing something artistic in printed woolen fabric, new statuary designs, and original ornaments, such as a new pattern in print or picture, painting, or an original configuration for an article of manufacture. The method by which the article is manufactured has nothing to do with the design or attractiveness of the article. Even the material used does not enter into the protection of the design of the product. For instance, a bust of a woman can be made of plastic, glass, metal, bronze, gold, or aluminum. Nevertheless, only one design patent covers all of them if the design is the same in each case. The process of making a bust, if new and original, can be a subject of a utility patent. Therefore, a product may be covered by both patents: a utility patent for making it and a design patent for the appearance of the product (if no prior patents exist).

There is a misconception that when a product is not patentable under the utility patent rule, because of prior art, the item can be patented under a design patent by incorporating certain features and marking the item "Patented." While certain advantage is obtained by such a scheme, the public is victimized by the practice because the word "Patented" leads the public to believe the idea that the product is protected by a utility patent, while it is not. Applications on design patents are strictly scrutinized by the Patent Office to avoid the granting of a design patent whose principal value arises from its utility. For that reason, articles that are protected by a design patent should be marked "Design Patent" and accompanied by the design-patent number.

The period of protection offered by a utility patent is 17 years. A design patent lasts for 3½ years, 7 years, or 14 years; it depends how fast the ornamental nature of the article is changing with the trend of the market. An inventor may patent his design for any of these three patent periods by paying an initial fee of $20 for filing, and a final fee of $10, $20, or $30, respectively for 3½, 7, and 14 years. An example of a design patent is shown in Fig. 11-1.

A design must be new and original and ornamental. In the statutes, the term *original* refers to inventive genius because the created design features are different from what already exist. Color and mere arrangement of the various design features do not necessarily constitute attractiveness and originality. Alteration of the material over the existing design patent does not constitute originality. Omission of a part or addition of some figures to the existing design to make it look distinctly different from the prior article cn make the design feature subject to patentability.

Design patents are subject to declaration of interference at the patent office when two inventors have filed their applications at the same time for substantially the same ornamental design. The procedure of determining the original inventor is about the same as for utility patents. For this reason, the originator of a design must file his design patent application before he displays his article in the market. This eliminates the risk of losing the invention to a competitor who—after seeing the invention—might file an application before the actual originator. A design should be kept secret until a patent has been issued on it.

REASONS FOR OBTAINING A DESIGN PATENT

An item with the same design used for a long time may cause boredom in the same manner that a person becomes tired of using

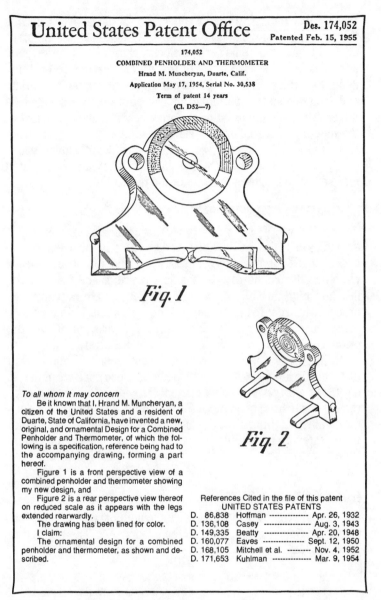

United States Patent Office

Des. 174,052
Patented Feb. 15, 1955

174,052

COMBINED PENHOLDER AND THERMOMETER

Hrand M. Muncheryan, Duarte, Calif.

Application May 17, 1954, Serial No. 30,538

Term of patent 14 years

(Cl. D52—7)

Fig. 1

Fig. 2

To all whom it may concern

Be it known that I, Hrand M. Muncheryan, a citizen of the United States and a resident of Duarte, State of California, have invented a new, original, and ornamental Design for a Combined Penholder and Thermometer, of which the following is a specification, reference being had to the accompanying drawing, forming a part hereof.

Figure 1 is a front perspective view of a combined penholder and thermometer showing my new design, and

Figure 2 is a rear perspective view thereof on reduced scale as it appears with the legs extended rearwardly.

The drawing has been lined for color.

I claim:

The ornamental design for a combined penholder and thermometer, as shown and described.

References Cited in the file of this patent
UNITED STATES PATENTS

D.	86,838	Hoffman	Apr. 26, 1932
D.	136,108	Casey	Aug. 3, 1943
D.	149,335	Beatty	Apr. 20, 1948
D.	160,077	Eaves	Sept. 12, 1950
D.	168,105	Mitchell et al.	Nov. 4, 1952
D.	171,653	Kuhlman	Mar. 9, 1954

Fig. 11-1. An example of a design patent.

one particular item year in and year out. Furthermore, if a utility device has been invented and the device has both utility and an ornamental value, two patents should be filed simultaneously. One patent will cover the usefulness and function of the patent and the other will cover the attractive appearance of the device (if such is the case). Fashions change not only in clothing, shoes, and hats, but also for household items such as clocks, china dishes, forks, pots, and pans, etc. Because many of these items cannot be patented under the utility patent statute, a design patent will be appropriate and desirable so that not only the market of the new design may increase but also the profit on the product may improve because of the meeting of the design to changing fashion and market trend. Because the cost of the design patent is considerably lower than a utility patent, changes in the design of an item can be instituted more often to heighten the sale of the product.

DOMINATION OF A DESIGN PATENT

As in a utility patent, a design patent dominates exactly what it embodies in its design drawing. If it dominates any article that is so similar that a purchaser might make a mistake in choosing one for the other, the patent may be considered invalid. The design of the new invention must be entirely different from what has already been produced in order to be a subject of a design patent. Unlike a utility patent, a design patent cannot be called an improvement over the prior art. This feature of the design patent is entirely distinct from the utility patent because a design invention should be original, new, and ornamental. If the article meets these characteristics, it has the property of becoming a patent (provided all other statutory requirements have been met). Furthermore, a design patent does not dominate any means for producing the design; the means will fall in the category of a utility patent. The design application can be superseded by a prior patent, the same as any utility patent, if the prior art is later than one year.

FILING A DESIGN PATENT

A design patent has the following constituent parts: A petition, a specification, a claim, an oath, and a drawing. The petition is

PETITION

To the Commissioner of Patents and Trademarks:

Petitioner,...................., a citizen of the United States and a resident of ...., State of ...., whose post office address is ...., prays that letters patent be granted to him for the new and original design for ...., set forth in the following specification.

Signed
_____
Inventor's Full Signature, Applicant

Fig. 11-2. An example of a petition.

```
                    SPECIFICATION

    To all whom it may concern:

    Be it known that I, . . . . . . . . . . . . . . . . . . . . , have invented a new, original, and
ornamental design for . . . . . . . . . . . . . . . . . . . . , of which the following is a
specification, reference being made to the accompanying drawing, forming a
part hereof.
    Fig. 1 is a plan view of a . . . . . . . . . . . . . . showing my new design.
    Fig. 2 is a side of the . . . . . . . . . . . . . . (if there is one).
    I claim:
    The ornamental design for a . . . . . . . . . . . . . . . . . . as shown and described.

                                        Signed
                                        _____
                                        Inventor's Full Signature
```

Fig. 11-3. An example of a specification.

```
                            OATH

State of . . . . . . . . . . .  )
                                ) ss:
Country of . . . . . . . . .  )
. . . . . . . . . . . . . . . , the above-named petitioner, being sworned (or affirmed)
deposes and says that he is a citizen of the United States and a resident of
County of . . . . . . . . . . . . . . . , State of . . . . . . . . . . , Zip Code . . . . . , that he verily
believes himself to be the original and sole inventor of the design for . . . . . . .
described and claimed in the foregoing specification; that he does not know and
does not believe that the same was ever known or used before his invention
thereof, or patented or described in any printed publication in any country before
his invention thereof, or more than one year prior to this application, or in public
use or on sale in the United States for more than one year to this application; that
said design has not been patented or made the subject of an inventor's certificate
in any country foreign to the United States on an application filed by him or his
legal representatives or assigns more than six months prior to this application;
that he acknowledges his duty to disclose information of which he is aware which
is material to the examination of this application, and that no application for
patent or inventor's certificate on said design has been filed by him or his
representatives or assigns in any country foreign to the United States, except as
follows (state anything that exists): . . . . . . . . . . . . . . .
                                        Signed
                                        _____
                                        (Inventor's Full Signature)

    Sworn to and subscribed before me this . . . . . day of . . . . . . . . . . 19 . . . . .
(Seal)
                                        Signed
                                        _____
                                        (Signature of Notary Public)
```

Fig. 11-4. An example of an oath accompanying an application.

```
The Commissioner of Patents and Trademarks
United States Patent and Trademark Office
Washington, D.C. 20231

Sir:

Transmitted herewith is the patent application of

Inventor: . . . . . . . . . . . . . . . . . . . . . . . .

For: . . . . . . . . . . . . . . . . . . . . . . . . . . .
    (Title of Invention)

Enclosed are:

One sheet of drawing.

Design patent application.

Filing Fee - $20.00
Fee for Years - · · · · · · · · · · ·
    Total - · · · · · · · · · ·
A check (or money order) in the amount of . . . . . . . . . . to cover the filing fee and
the final fee is herewith enclosed.
                                        Signed
                                        _____
                                        (Inventor's Full Signature)

                                        Address . . . . . . . . . . . . . . . . . . . . .

Date: . . . . . . . . . . . . .         . . . . . . . . . . . . . . . . . . . . .
```

Fig. 11-5. An example of an application transmittal letter.

similar to that for a utility patent. The specification is short, concise, and descriptive and it usually consists substantially of one paragraph. The claim is only one sentence. The oath is substantially the same as for a utility patent. The drawing of a design patent application should be drafted, if possible, by an expert draftsman or artist so that all new and original ornamental features of the invention are accurately and attractively delineated. If there exists any coloring in the ornamental features of the invention, the different colors may be shaded in accordance with the symbols given in Chapter 18.

Examples of a petition, a specification, a claim, and an oath are given in Figs. 11-2 through 11-5. The application should be accompanied by the prescribed fee of $20 for filing. When it becomes a patent, one of the fees indicated for the desired period must be paid.

Chapter 12
Evaluation
of a Patent

To realize commercial success, an invention must have commercial demand. To achieve this, the invention does not have to be a breakthrough or a basic invention. A great improvement over an existing product that is in demand by the public can give the patent a high value. A patent on such a product can become a *superinvention* when the inventor can secure a patent that cannot be infringed upon by competitors.

A claim that fully protects the invention described in the specification, as well as dominating all improvements and modifications that come thereafter, obviously will be more valuable than a claim that can be circumvented by minor changes. A patent with such a strong claim or claims will be respected and its value would be heightened to the maximum (provided all other business conditions have been taking their normal course). A strong patent will attract competitors seeking licenses from the patent owner because they would rather not attempt to circumvent the invention by producing inferior products (possibly) at a lower cost for sale at a reduced retail price.

While a patent grants the right to the patentee to exclude others from making, using, and selling the patented product, the government does nothing to protect the patentee from infringers. The only relief an inventor has is to bring a lawsuit against the infringer. When the inventor and the infringer are well funded to fight each other in a court, the superinvention with a strong patent posture will overcome any allegations that the infringer may advance in a court. He would be liable to all the legal costs, sales damages, and penalties set forth by the court. Only a few competitors, if any, will try to infringe a patent that their counsel has investigated and found it to be difficult for substitution.

To draft such strong claims, an inventor must be fully aware of any possibilities of circumvention by minor modifications of his invention. If modifications are possible, he should incorporate the modifications in his patent application. If necessary, more than one patent application should be filed to cover all modifications (species) that are possible. Any imprudent company, upon finding the financial stature of the patent owner to be weak, may risk the manufacture and merchandising of the patented product on the grounds that the patent owner will be unable to meet the expenditure for a lawsuit. In such an event, the patent owner should seek the financial aid of persons or firms who would be interested in his invention and will risk the outlay of funds necessary to pursue an infringement suit against the competitors.

HOW TO DETERMINE PATENT VALUE

In addition to ingenuity and skill, an inventor will need salesmanship and the ability to negotiate in order to convince a firm to risk financial support to pursue the practice of a patented invention. These efforts, together with the strong patent stature, will increase the commercial value and demand for the invention. Also, a useful patent has greater value in the hands of a person or corporation that has the proper financial backing than one that is very much interested in the invention and its commercialization, but lacks funds. There are many such "investors" willing to enter into a business connection with the patent owner. The patent owner will be much better off without the help or association of such parties because, in the long run, their association might be so detrimental to the success of the invention that the patent owner may find himself without a market for his invention.

How would an inventor know about such parties without first doing business with them? When these "interested" parties are very anxious to do business with the inventor and are promising a fantastic future for the commercialization of an invention, "of course for a small fee of several thousand dollars," that is time to stay away from them. Until an invention has been properly placed on the market, no one knows how successful the market for the product will be and how much the patent will earn for the originator.

An inventor could set a patent's earning capacity by estimating cost plus profit. However, the estimated earnings are not the value of the patent. Such earnings must be used only as an argument in the minds of prospective buyers and licensees. The value of an invention is influenced by the bargaining ability of the patent owner and his capability to practice his own invention prior to placing it before prospective buyers. Because of the amount of financial risk involved in the working of a new invention, the person, firm, or manufacturer that is approached will inevitably demand a major interest in the invention and patent. The value of a patented invention can be based on the estimated earnings or on a percentage of the earnings when the item is commercialized. The inventor should not be in a hurry to accept the first offer that comes along his way, but he should not be so stubborn in his demands that no purchaser will be willing to bargain with him for an equitable price. Terms are often for an outright purchase or on a royalty basis. Advance royalties are usually of prime consideration during negotiations.

METHODS FOR EVALUATING A PATENT

A patent is worth what an inventor can get for it. You should consider the novelty, the demand for the product, and the ease of manufacturing it. For these reasons, no manufacturer will jump at the opportunity of undertaking the manufacture and sale of an untried invention unless he can foresee profit that can be derived from the product. It is up to the inventor or patent owner to convince the manufacturer by displaying profit figures.

An accurate estimate of the potential for a product is very difficult to determine. If the nature of the invention permits, the inventor may produce, on a small scale, some of the products and try several retailers (using competent salesmen). The quantity may vary from several hundred to several thousand units depending on the type and cost of the product. On the other hand, if the patent owner is not financially able to support the manufacture of a small lot, then he must make an approximate estimate of the sales of his product.

An inventor who feels he has a superinvention covered by a superpatent should not have much difficulty in convincing others interested in his product. An inventor might also obtain opinions about the commercial value of his product by contacting various department store managers, retailers, and wholesalers doing business in similar lines. Having collected this information, he can make his own calculation on the basis of cost of production, distribution, discounts to wholesalers, and profit. Such an estimate in hand will be worth his while before approaching a prospective buyer of his invention. The most important part of his market search will be his determination of the value of his invention. Once this is determined and the inventor is equipped with information regarding the manufacturing and related costs, then he will be in a position to do any necessary bargaining.

Calculation of Cost and Profit

One method of evaluating the value of a patented product is to tabulate the following costs and profit as in this example:

Material Cost	$1.65	
Manufacturing Labor	3.15	
Overhead	0.78	
Distribution Cost	0.23	
Shipping Cost	0.46	
Packaging Cost	0.21	
Total Cost	6.48	
Manufacturer's Price	19.95	
Wholesaler's Discount	5.98	(30% discount)
	13.97	
Minus Production Cost	6.48	
Profit	$7.49	

In this example, the estimated profit is the minimum that an inventor can expect. This estimate is based on manufacturing a small product at low-quantity production. Some manufacturers determine their selling price by multiplying the basic cost by four for products that retail between $20 to $40. For higher manufacturing costs and higher retail prices, this ratio is reduced slightly.

If an inventor makes a deal with a manufacturer to produce and sell his product on a cash and royalty basis, then he can ask the manufacturer to give him a 10 percent royalty on the selling (wholesale) price of the product. There could be a provision that the manufacturer will sell a minimum of, for example, 25,000 units annually. The minimum royalty from the invention would be $34,925 per year. If the minimum—as specified in the agreement—is not attained, the patent rights revert to the inventor at his option. Such an agreement would be equitable to both parties because it would give the manufacturer an idea of the commercial value of the product. If the product is really in demand, the figures might be a hundred fold or more. This will give a sizable income to the inventor for his efforts and the expenditures he has undertaken prior to the sale. If the inventor wants to sell the product patent outright, he and the manufacturer should be able to come to an accurate conclusion.

It is difficult to sell a patent to a large company that has an engineering staff devoted to the development of new products. The company is reluctant to buy an invention with an unknown market background. It will be to the benefit of the inventor to first try small companies. In no case should the sale of the patent be carried out through a patent broker who promises high commercial return for the patent. Any patent broker who believes that a product is a money-maker will take the promotion of the sale or royalty stipulation on a contingency basis. A small financial outlay of not over $100 by the inventor to the broker for his preliminary paper work should be sufficient for office expenses. Anything above that amount would be unreasonable and it would indicate that the broker is not sure of the commercial value of the product.

The amount of money an inventor asks for his invention depends on how much he can get for it. His estimate of the market condition, demand for the product, the cost of production, and the volume of sale per year will give him a starting point to determine a sale price for his invention. It is more difficult to receive a large sum for an outright sale than if the invention is sold on a royalty basis with a certain amount of advance royalty or straight cash down. It would be easier for a small company to advance a reasonable amount of cash down and pay the inventor as the product is sold. If the inventor does not intend to market his own invention, this latter procedure might be his best opportunity for placing his product on the market.

Sale of Invention on Royalty Basis

For selling an invention on a royalty basis, a license is usually the predominant contract between the licensor and the licensee. In such an arrangement, the licensor (patent owner) permits the licensee to market his invention within certain conditions and monetary stipulations. The patent owner retains the title to the patent. The licensee agrees to make and sell the invention for a specified period of time. During that time the licensee makes periodic payments to the patent owner at a rate mutually agreed upon. If the license is made for the life of the patent it may be considered as a form of sale to the licensee.

There are two types of royalty licenses. An *exclusive license* permits only the licensee to make and sell the invention with the usual royalty payments. The second type is a *nonexclusive license*. Here the patent owner can grant permission to any number of licensees to make and sell his invention for specified royalties. In the latter case, the patentee also can manufacture and sell the invention. The royalty agreements need not be subject to the same terms and conditions with all licensees and the royalties do not have to be the same. The royalty and the conditions of license will depend on the stipulations made in the contract with each of the licensees. Granting of nonexclusive licenses by the patent owner sometimes is more advantageous. If one licensee fails to produce and sell the product in specified number, the sale from the other licensees might compensate for the loss. Furthermore, a broader commercialization is possible through the collective sale of the various licensees.

Exclusive License Basis

An exclusive license can have variations in its stipulations. For instance, the invention can be marketed by both the patent owner and the licensee. The license can give the licensee an exclusive right to commercialize the invention in a specified territory. The license grants the right for manufacturing and selling the invention for a limited time. The licensor can issue a license to a party, (in addition to the exclusive licensee) that has a business entirely different from the industry in which the exclusive licensee is marketing the invention.

SUMMARY OF PATENT EXPLOITATION

There are several ways an inventor can profit from his invention. The assessment of the merits of his patent is dependent on which of the options he has chosen to put in practice. A product has no commercial value until it is sold. It is easy to obtain a sympathetic expression from friends and associates, but it is difficult to get a fair commercial report strictly on the marketability of the product. Many large companies make a survey by sending free samples to various department stores or even to selected individuals. This procedure advertises the product and indicates if there is a demand for the product. This procedure, however, is not often within the financial capability of an average inventor. An alternative would be to sell the invention to a company engaged in a business in the same line or a similar product line.

A patent owner may sell his product on a royalty basis. He may license the invention on an exclusive license or an inexclusive license (the latter possibly being more beneficial to him financially).

He may also sell his patent outright or assign an interest in the patent to a second party.

In assigning any part of the patent to anyone, the inventor must be very careful in the preparation of the assignment. This is especially true when it is a partial assignment. Anyone having a partial interest in a patent may make and sell the product without accounting to the inventor or patent owner (if there is not specific stipulations in the contract). In assigning a part interest in a patent, there should be an agreement that states that neither party will manufacture and sell the invention without accounting the income from the sale. In such an arrangement both parties benefit. For instance, if the inventor of a patented product assigns 40 percent to another party for some monetary consideration and the other party manufactures and sells the product, then the inventor should receive 60 percent of the profits and the other party will receive 40 percent. The percentage of the profit interest might range anywhere from 5 percent to 95 percent. It depends on the parties entering into the contract.

When the patent owner (inventor) is well funded, the best method of commercializing the invention will be to manufacture the item himself and sell it through his company or distributors who have wide market facilities. A small inventor could start his own business and sell the product from his shop or by mail and gradually expand the business to a large-scale production. Many enterprising inventors choose the last procedure and do succeed in their new enterprises.

Any assignment in a patent should be registered in the patent office within three months after signing of the contract. If the assignee fails to have the assignment registered in the patent office, an imprudent inventor, within certain coverage of the law, may assign a number of other fractional interests to other parties and the assignee does not have real cause to sue the inventor for breach of contract because the contract was not registered in the patent office.

RECORDING OF ASSIGNMENTS

The patent statute provides for the sale of a patent or an application for patent by an instrument in writing which is known as *assignment.* Assignment can be the entire interest in the patent or a part interest. An assignment of a patent or a part thereof should be acknowledged before a notary public. If an assignment of the patent or an application is not recorded in the United States Patent and

Trademark Office within three months from its date of signing, the instrument becomes void against a subsequent purchaser without notice, unless it is recorded prior to the subsequent purchaser. (35 USC 261.)

An assignment of a patent application should identify the application by its serial number, date of filing, the name of the inventor, and the title of the invention. An assignment of a patent should indicate the patent number, date of issue, and the name of the patentee. If a patent application has been assigned and recorded before the date of issue of the patent, the patent will be issued to the assignee as an owner. In case the inventor and the assignee own the patent jointly, the patent will be issued to both as joint owners.

INVENTORS' RIGHTS IN AN ASSIGNMENT

When an inventor does not have sufficient funds to obtain a patent, he may call upon another party to finance his invention and patent application. In such an event, a contract for assignment should be written that is mutually satisfactory and equitable. The purchaser should determine whether the inventor owns the entire right of the patent or patent application. This precaution is necessary because if the inventor has assigned or licensed a part or all of his rights to another party then the purchaser must deal with both the inventor and the other party. If the inventor is unwilling to disclose any previous transactions with others, then the purchaser must abandon any dealing with the inventor. For this reason, an assignment search should be made by the purchaser prior to any purchase of the invention, patent, or the patent application.

The following cases illustrate the interpretation of the patent law Section 261, 35 USC, Patents:

Case No. 1. Let's suppose that Smith obtained a patent assignment from the inventor on March 1. One month later, the inventor imprudently assigned the patent to Anderson (who is considered an innocent purchaser). On May 1, Smith recorded his assignment in the patent office. When there is a court contest as to the ownership of the patent, Smith will prevail because he recorded his assignment within the three-month statutory period.

Case No. 2. In Case No. 1, if Anderson recorded his assignment in April—when he purchased it—Anderson would have prevailed because he recorded his assignment at the patent office.

Case No. 3. Smith obtains a patent assignment from the inventor on July 14. Four months later, on November 14, Smith records the assignment. But the inventor assigns the patent to

Anderson on September 14. Anderson does not record the assignment before Smith records his even though Smith is four months late. In a court action as to ownership, Smith wins.

Case No. 4. The inventor assigns his patent to Smith on January 1. Anderson obtains an assignment on the same invention on February 1 and records the assignment on February 15. On March 5, Smith records his assignment. When the two assignments are contested, Smith wins because he purchased the patent first and recorded the assignment within the statutory period of three months. If Anderson had recorded his assignment before Smith, still he would have lost the case because he is the second assignee. The case was fraudulent on the part of the inventor.

From the preceding examples, it should be clear that prior to the purchase of any patent or patent application the purchaser should write to the patent office to determine if any assignment on the invention exists. Also, the assignment should contain a clause stating that the inventor is the sole owner. In addition, if the assignment includes only a part of the patent, the inventor should see that a clause is included stating that the interest in the patent is "undivided."

ASSIGNMENT FORMS FOR RECORDING

The recorded assignment should contain the conditions and stipulations of the contract between the inventor and the purchaser. If it does not contain them it should have an identification reference to the assignment bearing the date of assignment. Figures 12-1 and 12-2 are examples of an assignment for a patent and an assignment for a patent application. Any other suitable form may be used.

INVENTORS' RECOURSE WHEN NO PATENT EXISTS

At times an inventor will submit an idea with a crude model to a company that is manufacturing analogous devices or products. The idea might be good, but it might be still in a premature operational state. The company might like the idea and might improve on it to make it a commercially salable product. The company might further exploit it to its own benefit without acquiring the rights from the originator (the inventor). Such an acquisition of an invention is known as *appropriation of invention.* Since the inventor does not have a patent and no written agreement with the company that has not solicited the submittal of the invention, the inventor has no legal recourse to filing a suit against the company to collect damages. In the first place, it is a grave mistake on the part of the inventor to

WHEREAS, I, ... , of the city of . . . . and state of . . . . ,
did obtain United States patent letters for an improvement in . . . . ,
No. . . . . , dated . . . . ; and whereas I am now the sole owner of
said patent; and,

WHEREAS, ... , city of . . . . , and state of . . . . ,
whose post office address is . . . . , is desirous of acquiring the entire
(or portion) interest in the same;

NOW, THEREFORE, in consideration of the sum of . . . . ($1.00) dollars,
the receipt of which is hereby acknowledged, and other good and valuable
considerations, I, ... , by these presents do sell, assign, and transfer
unto the said . . . . , the entire right, title, and interest in and to said
patent letters aforesaid; the same to be held and enjoyed by the said . . . .
. . . , for his own use and behoof, and for his legal representatives and assigns, to
the full end of the term for which the patent letters are granted, as fully and
entirely as the same would have been held by me had this assignment and sale
not been made.

Executed this . . . . day of . . . . , 19 . . . . . , at . . . .
State of . . . .)
) ss:
County of . . . .)

Before me personally appeared said . . . . and acknowledged the
foregoing instruments to be his free act and deed this . . . . . day of . . . . . 19 . . . . .

(Seal)   Signed ――――――――――――――
                                          Notary Public

Fig. 12-1. Assignment of a patent.

submit an invention to a company or an individual and to solicit
services or funding of the invention for commercial exploitation
without having at least a patent application.

To bring a law suit against any person or company for loss of
invention to that person or company is a very costly procedure.
Even any damages that the court might award to the inventor under
these conditions might not be sufficient to pay the attorney fees and
court costs. Consequently, an inventor should never attempt to
solicit any second party for financial aid in the development of a
working model or the manufacture of a perfected product until some
form of protection has been established for the invention.

A company that has received an unsolicited crude model of an
invention that later proves to be a workable product might feel that
it has expended a considerable amount of money to develop the idea
from almost nothing. The company might consider the invention as
its own product and go ahead on that basis to manufacture and sell it.
The only recourse an inventor might have in such a situation is to

apply for a patent. When the patent is issued to him, he can try to recover his invention by a legal suit against the company (after giving a notice of the patent to the company and waiting a reasonable time with no response from the company).

Law suits are costly. Therefore the inventor must evaluate the merits of the invention and its potential monetary value to him by suing and collecting damages or by going ahead and manufacturing it himself. Would such a legal effort pay his expenditure of action against the infringer? It is the inventor's decision at this point.

I know of a case when an inventor tried to stop a manufacturer from appropriating his invention, for which he had a strong patent position, and the patent was only a few months old. The manufacturer hired a lawyer who immediately declared the patent invalid

---

WHEREAS, I . . . . , of . . . . , have invented a certain new and useful improvements in . . . . for which an application for United States patent letters was filed on . . . . Serial No. . . . . (if the application has been prepared but not yet filed, then state "for which an application for United States patent letters was executed on . . . . " instead); and,

WHEREAS, ... of . . . . , whose post office address is . . . . is desirous of acquiring the entire (or portion thereof) right, title and interest in the same;

NOW, THEREFORE, in consideration of the sum of . . . . dollars ($ . . . .), the receipt whereof is hereby acknowledged, and other good and valuable consideration, I, the said . . . ., by these presents do sell, assign and transfer unto said . . . . , the full and exclusive right to said invention in the United States and the entire right, title, and interest in and to any and all patent letters that may be granted, therefore in, the United States.

I hereby authorize and request the commissioner of patents and trademarks to issue said letters patent to said . . . . as the assignee of the entire right, title, and interest in and to the same, for his sole use and behoof; and for the use and behoof of his legal representatives, to the full end of the term for which said patent letters may be granted, as fully and entirely as the same would have been held by me had this assignment and sale not been made.

Executed this . . . . day of . . . . 19 . . . . . at . . . .
State of . . . .)
) ss:
County of . . . .)

Before me personally appeared said . . . . and acknowledged the foregoing instruments to be his free act and deed this . . . . day of . . . . 19 . . . .

(Seal)   Signed _____
                                        Notary Public

Fig. 12-2. An example of an assignment of a patent application.

97

AGREEMENT

This agreement has been made by and between John Q. Public, resident of Anaheim, California, hereinafter to be referred to as the First Party, and Robert Q. Farmer, resident of Buena Park, California, hereinafter to be referred to as the Second Party.

WITNESSETH

WHEREAS, said First Party has developed and constructed a working model of an invention, known as Laser Printing System, and is desirous of filing a United States patent application in the Patent and Trademark Office and requires certain funds to defray the cost of filing and prosecuting said invention. The system generally consists of a source of laser beam that prints characters on paper, plastic, and metal at a cost comparable with conventional printing methods.

WHEREAS, said Second Party is desirous of aiding the First Party in securing a patent on said invention, and agrees to defray all costs arising from the preparation of the application, filing it in the United States Patent and Trademark Office, and prosecuting the application to issuance of a patent, the cost of which work is not to exceed $500.

It consideration of $500 in United States currency, said First Party agrees to assign 20 percent of the invention and patent, when issued, to said Second Party upon payment of said $500 to said First Party on signing this agreement.

The conditions and stipulations of this agreement between said First Party and said Second Party herein are as follows. The 20 percent interest assigned to said Second Party is an undivided interest of 100 percent of the invention and patent; that neither the First Party nor said Second Party herein will sell, transfer, assign or give any part of the said invention and patent to a third or fourth party without a written consent and agreement mutually drawn between the two parties herein; nor either party herein will make, use, or sell the said invention without sharing the income from said invention with the other party; that when said invention is commercialized, the First Party will receive 80 percent of the total revenue and said Second Party will receive 20 percent of the total revenue derived from said commercialization, said revenue will be held by both parties herein as if held by one person exclusively. When a patent is issued, the assignment shall be shown on the patent and the undivided shares of each party. In the event there is any expense in the commercialization of said invention, all such expenses will be deducted from the derived income prior to sharing the profits in the proportion delineated hereinabove. In case the invention is licensed to a third party, all monies, profits, and income from said invention and patent will be shared in accordance with each party's share.

This agreement is transferable to the heirs, legal representatives, or other legal assigns of the two parties herein upon the death of either party during the life of this agreement in the proportion as described and stipulated hereinabove.

Signed _____

State of California)   (First Party)

) ss:

County of Orange)   Signed _____

                                      (Second Party)

Sworn to and subscribed before me this . . . . . day of . . . . 19 . .

Signed _____

(SEAL)   (Notary Public)

Fig. 12-3. An example of an agreement.

without giving adequate reasons. His client continued manufacturing the product and deriving the financial benefit from it. Here was a case where the attorney had nothing to lose but everything to gain. In the event his client were driven to the court by the inventor, the attorney would collect his fees whether he won or lost the case.

The inventor took the matter into serious consideration and finally concluded that the financial rewards from the invention, even if he had won the case, and damages paid, the amount collected from the infringer for the few months during which his patent was infringed, would not recompense him for payment of attorney fees and the time lost in litigation. He dropped the case and the manufacturer continued to manufacture the product in competition with the inventor and patentee. How well does a patent protect an inventor's right to a product if he does not have the funds to pursue legal litigations to protect his right to the invention and patent?

Assume that while the inventor's patent was pending in the patent office a competing manufacturer also filed a patent application on the same type of invention. An interference will be called by the patent office to determine priority on the invention. Interferences sometimes take months or years before they are settled. In the meantime, neither party can stop the other from making and selling the product. For such circumstances that are not usually foreseen by the inventor, the inventor of a product should keep records of his invention even from the day of conception. He should also have his daily records witnessed by a confidential friend, or notarized from time to time until he applies for a patent. His record of invention development will be a prima facie evidence of possible priority if no one else can introduce any evidence that antedates the inventor's records.

Chapter 13
Protection of Inventions Without a Patent

Electromechanical and electronic inventions can be protected without securing a patent on them provided that the product to be protected is either permanently welded or its parts are so secured that to disassemble the device one must destructively take it apart. In such cases, it is almost impossible to duplicate the invention. When a patent is issued to an inventor, a tendency exists for others to infringe on the invention or to duplicate it with a slight change in the mechanism or parts of the device. The remedy to stop a person or a company from "stealing" the invention, even when it is patented, is to conceal the structure of the invention from the public even though the procedure retards the advancement of the art in the particular line of invention. Patenting an invention is for the purpose of preventing others from making, using, and selling the patented invention. If an inventor is financially unable to prevent the unauthorized commercialization of his patented product, there must be some other method of concealing his invention (generally) from the public at large because patenting the invention places it before the public when it is published in the *Official Gazette* and in the patent letters.

While I am reluctant to advocate the use of concealment as a means of monopolizing a product of mental ingenuity, the present business conditions and our patent laws do not protect the inventor against having his patented invention jeopardized by others or by their making and selling the patented invention without the consent of the inventor. The remedy for this is to bring a lawsuit by the inventor against infringers. But if the inventor is not financially able to meet the high cost of a lawsuit, what alternative would an inventor have? Imprudent competitors, knowing that the inventor cannot fight them back to support his right of patented invention, will practice the invention without regard to the inventor. It is being done every day.

How can an inventor protect a mechanical invention having so many moving parts geared together in an assembly that can be disassembled step by step to determine what parts constitute the structure of the device and how they are arranged to cooperatively work together? Any invention that can be disassembled and examined visually cannot be protected by the manner to be described in this chapter. The only remedy is to obtain a good, strong patent on it and to take the risk that no one will attempt to infringe on the invention. Furthermore, when an infringement has been committed and the infringer does not terminate his infringement activities—upon formal notification by the inventor of the existence of his valid patent—you should try to find an interested person who will advance sufficient funds (for a part of the patent right) to bring a suit against the infringer. Most mechanical patents cannot be protected by the manner described in this chapter. They should be patented.

ELECTRONIC INVENTIONS

Some types of electronic inventions will be ideal for protection against competitors by the use of the procedure described in this chapter. For instance, most electronic products consist of resistors, capacitors, diodes, transistors, or integrated circuits deposited in silicon or sapphire wafers of very small configurations. Because there will be no movement in these components, they can be assembled on a breadboard, bonded or soldered together in the usual way for the particular circuit, and then gently squeezed together to a small mass (see Figs. 13-1 and 13-2)

As shown in Fig. 13-1, the component parts are connected by means of leads or wires, that separate the components one from the other so that the final assembly occupies a relatively large area. In Fig. 13-2, the same circuit is compressed by squeezing the leads together to bring the parts closer to each other. Now take a polyethylene sheet of about 1/32 of an inch thick and form a small boxlike container to hold the circuit parts. Tape the sides so that when a liquid potting compound is poured on the circuit the liquid will not ooze to the exterior.

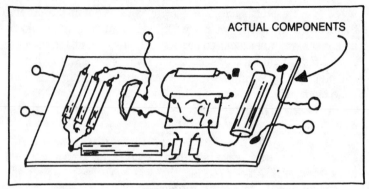

Fig. 13-1. An electronic circuit assembly.

Prepare a mixture of dark-colored resin, as that sold by Epoxylite Corporation, Formula 805/383, or Castex 202 of Industrial Chemical Specialties Corporation. Mix the component parts in the proportion recommended by the manufacturer and pour it on the circuit contained in the polyethylene container. If the curing time, for example, is 1 minute, wait about 30 to 40 seconds after pouring and insert several resistors or diodes (that have been discarded because of their previous failure) into the various sections of the curing mass of resin. Be careful not to touch the wiring of the other components in the circuit. These latter components may stick out from the surface of the resin, and they are used to camouflage the invention. Have the external leads of the circuit project from the sides of the polyethylene container and mark them for your own information. When the resin cures, you have a mass of plastic material with a few electronic components and the leads extending out from the cured mass. Remove the polyethylene container by splitting it with a knife if necessary. The finished product will have input-output leads, the resinous mass, and the camouflaging components. No one will know what has been potted in the resinous mass.

Why can't the resinous mass be dissolved in a solvent to remove the circuit intact? Such potting of electronic circuits have been done and are being done every day by various electronic companies to save space in their circuit assemblies. Furthermore, there are solvents that dissolve the resin so that the circuit assembly can be recovered for examination during a failure analysis. There are resins that are used for forming molds for castings, these resins are extremely hard. When cured, the potting compound cannot be dissolved in a solvent the same as bakelite. The only way to remove the circuit from the potting mass is to drill out the potting compound. This procedure will destroy the circuit connections or the component parts, and make the procedure for uncovering the components intact useless. I have routinely produced electronic products for consumer use by using the procedure outlined above. The main purpose of the potting procedure is to make tampering with the manufactured products impossible even though the products are covered by several United States patents.

When a patent is issued, the information in it becomes public knowledge and might instigate imprudent persons to try to duplicate the product by following the description in the specification. Someone might even modify the product before placing it in the market. When such a person has no knowledge of the construction of the product, then the inventor will have the protection that no patent can offer.

CHEMICAL INVENTIONS

Chemical inventions consist of such things as medical preparations, household cleaners, washing compounds, polishers, soaps, plating solutions, metal surface treating compounds, photographic chemicals, fuels, paints, pyrotechnics, textile treatment chemicals, and similar materials. While most compounds can be analyzed chemically or spectrographically—especially when the materials are soluble in aqueous solvents or are ionizable for spectral work—there are certain organic materials that dissolve or mix with a new chemical formula and do not ionize. These substances are not

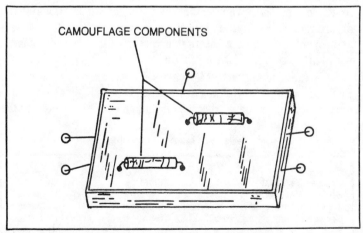

Fig. 13-2. An example of an electronic circuit in potted form.

vulnerable to analysis that can give a significant result. Because most organic compounds consist of carbon, hydrogen, oxygen, nitrogen, and inorganic elements in some cases, it is difficult to determine the radical or the compound that the organic elements are derived from in the spectrochemical analysis. Plain chemical analysis can be used, but in complex mixtures the derivation of the elements analyzed cannot be always determined accurately.

In chemical mixtures, the elements or compounds entering into the combination do not always dissolve. They can remain chemically separate. This characteristic of the mixture can aid in the analysis by simplifying separation during analysis. It is easier to camouflage a mixture than a solution in which a number of different compounds have been dissolved and ionized.

When a patent is issued on a new and useful compound, the composition of the new formula is open to the public. For instance, suppose laundry soap builder, No. 1 (see Table 13-1), is new and a patent has been granted because no prior art exists. Let's further assume that the inventor patented the combination shown in No. 1, and failed to state a range for each of the chemical compounds present. Then a second party sees the formula in the patent and combines the same chemicals in the proportions given in No. 2 (see Table 13-1). If he wants to obtain a patent and no prior art exists on a combination as shown, then he is granted a patent on formula No. 2. Actually, the second inventor is using the first inventor's idea. However, he claims in his application that his combination yields more suds in a soap mixture and cleans the washed material better. Therefore he claims an improvement over the first invention. If the first inventor had given a range of possible combinations of the different compounds, then it would have been difficult for the second inventor to imitate the first inventor's formula and compete with him in commercializing the product.

On the assumption that formula No. 1 was the only combination that was possible in yielding a substantially satisfactory soap builder, the second inventor then could camouflage the formula by the

Table 13-1. Laundry Soap Builder.

	No. 1	No. 2
Sodium Carbonate	22 grams	25 grams
Sodium Metasilicate	13 grams	11 grams
Trisodium Phosphate	40 grams	44 grams
Sodium Bicarbonate	15 grams	12 grams
Tetrasodium Pyrophosphate	10 grams	8 grams

Table 13-2. Electroless Plating Solution.

	No. 1	No. 2
Silver Nitrate	25 grams	25 grams
Ammonium Chloride	11 grams	10 grams
Sodium Thiosulphate	25 grams	25 grams
Whiting	225 grams	200 grams
Tartaric Acid	13 grams	13 grams
Sucrose	—	10 grams
Water	480 milliliters	500 milliliters

use of a chemical material that has no chemical function in the new combination. Because the chemical materials used in formula No. 1 are soluble in water as well as ionizable, they can be easily analyzed and the camouflage determined as adding nothing to the capability of the formula for the intended use. The inventor then can bring a lawsuit against the second inventor if he is financially able to do so.

Suppose another inventor formulated a mixture for plating materials with silver using an electroless method with formula No. 1 shown in Table 13-2. Formula No. 1 must have its component ingredients in the proportion shown in order to decompose the silver nitrate without contaminants in the decomposed silver metal. The presence of whiting does not enter into any chemical combination because it is pulverized chalk that does not dissolve in water. Therefore, the amount of whiting can be varied without affecting the chemical reaction of the other compounds with each other. Ammonium chloride, tartaric acid, and thiosulphate decompose the silver ions in the solution into metallic silver on a metal such as copper or brass. Ammonium content is reduced by 1 gram in formula No. 2 (Table 13-2). This reduction is compensated for by the presence of sucrose in the solution. Sucrose is used as camouflage because it does not ionize in water. Nevertheless, in the presence of an acid, such as tartaric acid or hydrochloric acid from the chloride ions in the solution, the sucrose can become hydrolyzed and it can affect the decomposition of the silver ions into metallic deposits. It is very difficult to analyze sucrose in the presence of other inorganic compounds. Here is a solution that does not permit ready analysis. Therefore the inventor can use it without regard to a patent if the formula No. 1 did not exist. If formula No. 1 were patented, the first inventor would not be in a position to press charges of infringement because he could not prove the exact composition of formula No. 2. Although the preceding discussion is purely hypothetical, the substance of the procedure can be put into practice without infringing on anyone's patented product provided, as emphasized, the chemical constituents added do not enter into a chemical reaction or become reactive due to the presence of hydrolytic substances in the solution.

LASER INVENTIONS

Laser generators cannot be protected in the manner described for electronic circuits. Nevertheless, the circuits contained in portable laser equipment can be potted in the manner described under the heading of "electronic inventions." Most laser systems or devices are destined to be used by government agencies, hospitals, private institutions, and research laboratories. It would be very difficult for a patent owner to become aware of the unauthorized use of his invention by other manufacturers. His only recourse would be to install "pottable" circuits in his device or system.

LOSS OF A PATENTED PRODUCT TO IMPRUDENT FIRMS

This situation can be more thoroughly described by giving examples. Suppose an inventor has a superinvention covered with a superpatent so that circumvention is very difficult if not impossible. The XYZ Corporation has been furnishing the government with systems and devices of their proprietary products. This corporation, always searching for new products along its manufacturing line, comes across the inventor's patent in the *Patent Gazette*. The company purchases one or more copies of the patent and has its engineering staff study the patent. They find the device superior to the proprietary product they are selling to a particular government agency. They decide to use the patented device in their systems, and they use and continue selling it to the government.

The inventor has no way whatsoever to be aware of this condition because most of the equipment, systems, devices or other items sold to the government are classified as confidential or secret. Neither the manufacturer nor the government agency can give any information regarding any such products being procured by them. The manufacturer who has the government procurement contract reaps the benefit of the inventor's mental ingenuity and proprietary product without regard to the inventor. In rare cases the manufacturer is very honest and will inform the inventor or the patent owner of the company's interest in the patented product and will make an offer. If the inventor does not agree to the offer made by the manufacturer, he may lose his chances of making a deal with the manufacturer. On the other hand, if he cooperates with the manufacturer he might even receive an offer to join the firm to pursue the development and further modification of the device.

Another example is a manufacturer of medical products that sells a product to various hospitals and medical institutions. This is a difficult area for the inventor to check for manufacturers' activity. A manufacturer, having seen the inventor's product in the *Patent Gazette*, incorporates the product in his system and sells it to the hospitals, dental product houses, and other medical firms without the knowledge or consent of the patent owner. This manufacturer considers the invention to be his own because he incorporated it in the system which he sells. Naturally, the inventor has no knowledge of his invention being exploited by the manufacturing firm. What can the inventor do? Nothing except to go ahead and make his invention, if he is financially able, and commercialize it in competition with the first manufacturer. His compettitor probably already has the major segment of the market.

I wish that these examples were only hypothetical, but they are not. Such things happen every day in any country where a patent system exists. There are a few exceptions where the government of the country has so systematized the patent laws and regulations that it would be considered a criminal offense to "steal" anyone's patented invention. We do not have such a law in the United States. The patent owner is responsible for protecting his proprietary rights by resorting to the federal courts, if necessary, to exclude others from manufacturing and selling his patented product (if he is aware of others who are manufacturing and commercializing his product). In addition, he must have the necessary funds to finance the court costs. Such a procedure is not usually within the financial means of a small inventor.

There are cases where an innocent inventor has lost his patented invention to others who reap large profits from the exploitation of his invention. For instance, a patent broker advertises in popular magazines, which many inventors read, stating that he will pay cash for useful and meritorious inventions. He might promise that he will develop, if necessary, and take out patents on the device before he presents it to "anxious manufacturers who are waiting to buy the inventor's product." The only catch is, that the inventor will have to advance a small fee $1000 or $2000 for all the work the patent broker will perform. The inventor shouldn't mind because the invention will no doubt bring in thousands or millions of dollars and make the inventor rich almost overnight. In the opinion of one of the most reliable and competent firms in the United States that helps inventors to place their inventions on the market, Kessler Corporation, these offers are rip-offs and the inventor should never go near promoters who promise to make someone a millionaire in a very short time.

There is another group of patent brokers who advertise that they have connections in Hong Kong or in Korea where they can have the inventor's new product manufactured at a very low cost and even distributed by them at no cost to the inventor. All they want is the inventor's trust in them and his invention—patented or unpatented. The inventor happily complies with their request and sends his model, drawings, description or any other material to help the manufacturer to understand the construction of the invention. With honest brokers this is an ideal situation. But with dishonest brokers, it is anathema to the invention from the standpoint of the inventor. He never hears from them. When he writes to them (to a post office box, of course) his letter is returned with the envelope marked "Addressee moved with no forwarding address left."

One such broker, operating from a northern California post office box address, received information from an inventor on his product and apparently sent the informative file to his Hong Kong associates. The inventor heard from the Hong Kong firm stating his invention was "very good and acceptable" and that the inventor would hear from them shortly with regard to contract stipulations, etc. The inventor never heard from them again even after writing to them several times.

Four years passed, the inventor's product appeared on the market in the United States. The inventor bought a sample from the distributor. The product was packaged in a cardboard box having six color illustrations on it together with a marking stating "Patent Pending." There was no manufacturer's name or address on the package. Inside the package, at one inconspicuous point, was a mark indicating that the product was made outside of the United States. The inventor's attorney tried to trace the sellers or distributors without success.

The patent office was informed about the apparently fraudulent use of "Patent Pending." The office turned a cold shoulder even though the patent law states a fine of $500 will be instituted against the user of the false marking. The law further states that $250 will be paid to the reporter and the other $250 will go to the government. The case has died away and the infringers are reaping the rewards that should have been enjoyed by the inventor.

There are countless cases of fraud and exploitation of the inventions. On the other hand, there are many honest invention licensing firms of high standing that are helping inventors locate commercial outlets for their inventions. All one can do is to exercise caution in selecting the right type of invention negotiating and licensing firm. I suggest that any inventor who is not marketing his own invention and wants to have an agency to handle the exploitation of his invention to write to one of the following firms for advice and assistance: Kessler Sales Corporation, 1247 Napoleon Street, Fremont, Ohio 43420 or University Patents, Inc., 875 N. Michigan Avenue, Chicago, Illinois, 60611.

HOW TO FIND AN HONEST PATENT NEGOTIATOR

When considering a company or patent agency to handle the commercial promotion of your patented invention, request that the company send references. Contact the chamber of commerce of the city where the business is located. You could also contact Better Business Bureaus and bankers with whom the company is dealing. Even after having the references, be sure you do not pay more than $100 or $200 for the preliminary paperwork and office overhead. Try to arrange your business contract on a contingency basis, if possible, or on some percentage of the profits or royalties received. Competent patent negotiators do have good contacts with various firms of the industry and they are skilled in transacting business in connection with inventions. All you do is to be cautious in your selection. In the long run, you will be happier to be associated with an honest and competent firm.

Chapter 14
Product
Development

At present the trend to spend more money toward research and product development is increasingly becoming important. In like manner, the trend toward employment of a cooperative teamwork is also becoming necessary. In a teamwork, each individual trained in his line of employment contributes his specialty to the conduct of the research, whereby emerges a new life idea which gives rise to the development of the idea into a working model. The model is improved by additional ideas during its operation and turns into a superior product over the prior art. The improvement, then, is patented to give it the distinction of a proprietary product. Through this effort, together with well-funded support from the governing body of the teamwork, a superinvention finds its way into the market through a successful commercialization program. Then what about an individual who is working by himself with moderate funds? What does he realize for his "brain child?" What are his chances of success? Let us take the product development in steps to see where the individual inventor will find himself in this maze of complex financial world.

PATENT SEARCH PRIOR TO DEVELOPMENT

An individual with a good idea wants to develop it, patent it, and open a successful market for it. He has the idea but is not quite sure how to put it to work. Before attempting to patent the product, he must make a preliminary search for prior art. He makes a patent search in a public library where a patent section exists. He consults with the patent librarian who finds various patents in the same classification as the inventor's idea. The inventor studies them and finds several methods of making his idea work. He has now learned how others have developed a similar idea into working products and the degree of success with which the product is operating. His product has to be an improvement over the existing art as well as novel and workable so that he can secure a patent for it.

He begins to improve on the patents he has studied. Before long he finds his efforts are developing into a new and useful invention. He compares his invention with the prior patents to determine if any part of his invention has been superseded by prior patents. If there is any superseding art he tries to avoid a conflicting part or step by eliminating it or making a substitution. He then studies his entire idea to see if the innovation is operating as expected. He makes a model and tries it. He finds not only that it is operating as it should, but that it has one or more advantages over the existing patented products.

The first thing he undertakes is to make a drawing of the working model and he writes a description of its construction and operation. He thinks of applying for a patent. Because he is a conservative person and very serious about his undertaking, he wants to make further search through books and periodicals in the library to determine if anything newer than the issued patents exists on the product. Now that he has become expert on the development and construction of the product, he wants to make it a superinvention so that circumvention by competitors will be difficult if not impossible. His next step is to search the market for other similar products.

The larger the number of variations of the invention (species) that the inventor may conceive the greater is the likelihood of drafting broad and inclusive claims that will be effective for the duration of the patent. If fewer of these variations are conceived, and there might not exist many in the particular case, the inventor still has the advantage over any future competitor who may want to circumvent his invention. Furthermore, if the inventor cannot conceive many variations prior to the application for a patent, but such new ideas might come to him after the filing of the patent, then he can still file the additional application covering the newly conceived ideas. The inventor may include in his application as many as five different species (variations) of the invention. If the number of species exceed five, and he must file a new application to cover the additional species over five. The second application will have the benefit of the date of the first application if the latter has not become a patent at the time of second filing.

Let's assume that while the applicant's (inventor's) application was still pending, he discovered a product being sold on the market that is analogous to his invention. The product is not labeled "patent applied for," "patent pending," or "patented." The inventor has legal right to amend his broad claims and making them broader by covering the essential features of the competitor's product not covered by the inventor's broad claims. The inventor's action is perfectly honest and legal because apparently the competitor did not have a patent or application pending because he did not label his product. When the patent is issued on the inventor's product, the patent will cover the competitor's product as well as his own invention. In such an event, the competitor has to obtain a license from the inventor to use his patented invention. Otherwise he would be infringing on the inventor's patent.

Such a situation is indicative of the importance of applying for a patent. Because the inventor will be filing his own patent application, which will cost about $150 for filing (saving hundreds of dollars by not using the services of an attorney), he can afford to apply for a patent as soon as he thinks his invention has been completed (if it avoids prior art in accordance with the result of this search). He can later file additional patent applications as he conceives additional variations of his invention (species). The additional applications, when allowed, will strengthen his patent position.

Some inventors have the notion that if they file or register a certificate of disclosure at the patent office, the action will protect them against any competitive business or infringers. This is inaccurate. The certificate of disclosure, as mentioned in the oath, is worthless from the standpoint of protecting the invention. Some imprudent patent brokers try to make the inventor believe that such a measure will protect the invention. They charge a fee for furnishing blank copies of the certificate of disclosure to the inventor. Copies of these forms are obtainable free of charge from the patent office by writing to the attention of the commissioner (the address is given in Chapter 1 of this book).

When the certificate is filled in accordance with the instructions printed on the form, the inventor returns two copies together with a fee of $10 for registration at the patent office. The office records it and sends a copy back to the inventor with the patent office seal affixed to it. This certificate is good for two years. Good for what? It is useless until the inventor applies for a patent within a reasonable time after registering the certificate. The certificate of disclosure is good only when the inventor files a patent application within two years and when another applicant has also filed for a similar invention and priority question is to be determined (in an interference case). If the filing date of the certificate is earlier than the conception or filing date of the second inventor, it helps to resolve the priority of the invention. If the inventor has waited two years before filing, he has to prove that he was diligent in improving his invention during the period before filing. Accordingly, it will be best to file for a patent application as soon as it is feasible after registering the certificate of disclosure (or even sooner).

MARKET SEARCH FOR PRIOR PRODUCTS

An inventor visits department stores, equipment accessory stores, wholesalers and retail stores inquiring whether they are selling a product similar to his. A number of stores he has visited do not display any product that resembles his invention. It is a very happy feeling. One store, however, happens to be selling something similar to his invention. He purchases a sample, goes home, and disassembles it. He finds that the disassembled model does not contain the parts and steps he has incorporated in his invention. Everything so far goes very smoothly and he is now seriously thinking of converting his rough sketches and description into an official patent application. He prepares his patent application and files it with reasonable assurance that the patent office will not come up with additional patents or prior art that will anticipate his invention.

After six to eight months he hears from the patent office. That is his first official action from the patent office. The examiner has brought up a few objections or questions concerning the parts of the drawing or specification. The examiner might even have rejected one or two claims for indefiniteness or on prior art that he cites and includes the copies of patents, if any, in the same envelope of the official action. Upon studying the examiner's action, the inventor notices several of the patents cited are those he had already seen in the library and he had circumvented them. He is right in that he has circumvented them. Why does the examiner cite them against his invention?

The examiner cites them because he is not aware that the applicant has already seen them, and it is his duty to uncover any prior art that has bearing on the inventor's product. The examiner gives the inventor an opportunity to compare his invention with those shown in the prior art. It is possible that the inventor missed one single point that would be important for him to cover in his claim or claims. At times, the examiner also makes suggestions to the inventor regarding the steps to be taken to improve on his claims so that they fully cover his invention. There has been a time when an examiner has written an exemplary claim when he has found that there is sufficient merit in the invention to make the claims as strong as possible.

The inventor seizes the opportunity given by the examiner by his objections or rejections and amends his claim or claims in the light of the examiner's instructions. His claims now stand more secure and less circumventable by possible competitors. The inventor has started with a mere idea, he has converted it into a working product by examining and studying prior art, and his application has been critically studied by an examiner for any errors in the coverage of his invention for a strong patent.

It can be readily seen that the development of a patent application is a continuous process of changes and improvements in the product aided by the existing art and by the examiner's disclosure of prior art. A common belief is that when the examiner does not like the idea or the manner it is represented in an application form he immediately rejects. This is a fallacy. He rejects something on the grounds that the applicant has made errors in the application or the claims are weak and he thinks stronger claims can be obtained (although he does not state this fact to the applicant). He gives the applicant the opportunity to amend his work for the issuance of a strong patent complete with pertinent claims. It must be kept in mind that if the examiner rejects an application or a part thereof on certain grounds and the applicant has not amended the application, the application does not become a patent. When an application is turned down because of errors or prior art, the patent office does not receive final fee and this is detrimental to the maintenance of the patent department. The examiner would be in favor of accepting a patent application and allowing it even in the first action if he finds any allowable matter in the application so that a valid patent can be issued.

COMPLEMENTARY DEVELOPMENT OF THE PRODUCT

The efforts of the applicant—through prior art search, the amendment, or changes made by the applicant during manufacturing of the article—have now finally improved the invention so much that it might require an additional patent to fully cover the original product together with the improvement made on it. Such patents, when granted, are called *fortifying patents*. These patents are useful in the prevention of circumvention by the competitors. The product then can be marked "Patented, Other Patents Pending." This marking will announce to others that the product is fully covered by patent, and in most cases will exclude any competitor from attempting to produce any product resembling any part of the patented invention. Product development does not stop after a patent is granted; it is a continuous process of improvement and protection by a series of patents.

A prudent manufacturer of a good product never stops improvement on it. These improvements may arise as a result of new ideas by his engineering staff or by making changes on the structure of the product to reduce the cost of manufacturing, and consequently the selling price of the product to increase sales. Likewise, when an inventor has licensed his invention to a manufacturer on a royalty basis, he should not stop improvement on his invention (if this is possible). Some license contracts include a clause stating that any improvement on the product by the inventor will be a part of the invention with or without further compensation for the inventor's improvement. The inventor should make an attempt to include in the contract a compensation clause on any new improvement, with the patenting cost to be defrayed by the licensee. If such a clause is not forthcoming in a license stipulation, the situation could eliminate the incentive by the inventor to make further improvements on his invention. On the other hand, the stipulation should also include a clause so that when the manufacturer makes any improvement the inventor will benefit from such improvement as if it has been contributed by him.

Chapter 15
Trademark Registration

A *trademark* is a legally reserved, distinctive name, symbol, or mark placed on a commodity so that the buyer of the commodity can distinguish the owner's product from other products. A *trade name* is often related to the company's entire business and good will. Unlike a patent, the proprietary right borne by a trademark is perpetual as long as the company is in business.

This does not signify that the purchaser of a product bearing a certain trademark must know the merchandiser personally. The mark distinguishes the product that has been tried and found by the buyer to be desirable over other products. The buyer can rely on the product identified by its trade name that what he is buying is of the same quality he is acquainted having used it in the past. Thus, a trademark identifies the product while a trade name indicates the origin and the manufacturer of the product. For instance, UNEEDA is a trademark for a certain type of biscuit manufactured by the National Biscuit Company, and NABISCO is the trade name of the same company. In other words, The National Biscuit Company or NABISCO manufactures various biscuits trademarked by RITZ, NILLA, ROYAL LUNCH, etc. Through advertising, these marks guarantee the quality of goods and arouse and sustain demand for the goods. All these goods are collectively known as NABISCO products.

The right to a trademark must be created by its constant use in both interstate and international commerce. Trademarks are not protected by patent or copyright laws. To be called a trademark, the mark must appear on the commodities wherever the merchandise is sold.

If Company A uses a mark that is confusingly similar to the trademark registered by Company B, and stamps the mark on products that are entirely different from those of the trademark owner (Company B), Company A is not considered to be infringing on the trademark of the registered mark owner. A trademark is held similar to another mark if the similarities of the two marks are so close that the general public can confuse one for the other. For instance, a lion's head imprinted on certain goods of a company can be confusing to persons who read the words "LION'S HEAD" on the goods of another company. While in the past such marks were applied on different companies' goods (whether they were of the same kind or not), the present trend is toward restricting the trademarks to marks or words that do not bear similarities or confusing connotations.

TYPES OF TRADEMARK REGISTRATIONS

In order to monopolize on a trademark, it must be registered in the United States Patent and Trademark Office. Any mark that is not registered is public property and anyone can use it. The registration of a trademark means that prior art has been searched by the patent office, and novelty and related statutory requirements have been met. Therefore, any unauthorized use of the registered trademark constitutes infringement.

The United States Trademark Act, and as amended thereafter, provides two types of trademark registrations: *principal registration* and *supplemental registration* or *register*. Under the United States Constitution, every state may provide registration for a trademark to be effective only within the state where it is registered, and its administration is regulated by the state laws. When the mark is being used in interstate commerce and foreign trade, the registration and its control are the functions of the federal government. A trademark should be registered with the federal government in order for the owner to have use of the mark throughout territories of the United States. This is especially important when there is a likelihood of business expansion in the future.

Principal Register

The following are examples of registrable trademark goods.

☐ Manufactured articles such as clocks, lamps, toasters, rehis prior use of the mark on goods sold between states or in foreign commerce. As in patent rights, the owner of the trademark can sue any infringer in United States courts.

A trademark cannot be descriptive of goods on which it is being affixed. For example, the manufacturer of vanilla biscuits cannot obtain a trademark that states "Vanilla Biscuits." Any manufacturer is free to produce vanilla biscuits and mark his merchandise "Vanilla Biscuits" so that customers will know the content of the package. However, such a manufacturer may mark his vanilla biscuit boxes with a coined name such as "Golden Circles." When registered, this trademark belongs only to one company. No other singular company may mark biscuits "Golden Circles."

Suggestive marks cannot be registered under the principal register. Examples of such marks are "Peerless Cleaner" for a soap trademark, "Emolient Cream" for cold cream, "Noncholesterol Peanuts" for peanuts canned in boxes or bottles, "Sharp-Edge Razor Blades" for safety razor blades, "Dark Writer" for a pencil, "india ink Stylus" for a pen using india ink, etc. Nevertheless, "Camay" may be used for a certain manufacturer's soap", "Peans" can be used for peanuts (because it carries no meaning except referring to the particular manufacturer's product), and "Mark II" could be used for razor blades even though that name is used on a certain automobile. The latter example is true because the two products are two different types of commodities and there exists no confusion between the two products. A technical mark to be registered must be in use for at least five years on goods sold by the manufacturer.

Supplemental Register

Trademarks that do not qualify for registration on the Principal Register, because they lack distinctiveness are registrable, on the Supplemental Register. These marks must be in use for at least one year in interstate commerce or in foreign trade. A geographical trademark, a descriptive trademark, or a proper name, through long usage, may become distinctive and registrable on either a Principal Register or on a Supplemental Register. Examples of such names are *India* for ink, *Perfect Circle* for piston rings, and *Gillete* for razor blades or the razor itself. Any trade name that may be considered descriptive during its early years may be registered on the supplemental register. Through continuous use of the trade name, it can become distinctive. Then it may be registered on the Principal Register. Another advantage of the Supplemental Register is that a confusion between the two products. A technical mark to be registered must be in use for at least five years on goods sold by the manufacturer.

Supplemental Register

Trademarks that do not qualify for registration on the Principal Register, because they lack distinctiveness are registrable, on the Supplemental Register. These marks must be in use for at least one year in interstate commerce or in foreign trade. A geographical trademark, a descriptive trademark, or a proper name, through long usage, may become distinctive and registrable on either a Principal Register or on a Supplemental Register. Examples of such names are *India* for ink, *Perfect Circle* for piston rings, and *Gillete* for razor blades or the razor itself. Any trade name that may be considered descriptive during its early years may be registered on the supplemental register. Through continuous use of the trade name, it can become distinctive. Then it may be registered on the Principal Register. Another advantage of the Supplemental Register is that a search through the patent office can be made prior to the registration of the mark. It also permits the registration to be obtained in foreign countries after it is registered in the United States.

PARTS OF A TRADEMARK APPLICATION

A trademark registration application comprises a drawing, a specification and a claim, five specimens of the trademark, the name and the address of the registrant, and the filing fee.

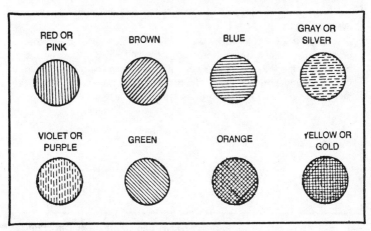

Fig. 15-1. Designations for various colors used on drawings.

```
                                    Mark . . . . . . . . . . . . . . . . . . . .
                                       (Identification of Mark)
                                    Class No. . . . . . . . . . . . . . .
To the Commissioner of Patents and Trademarks:

                    . . . . . . . . . . . . . . . . . . . .
                         (Name of Applicant)


                    . . . . . . . . . . . . . . . . . . .
                      (Full Address of Applicant)


                    . . . . . . . . . . . . . . . . . . .
                      (Citizenship of Applicant)
```

The above-identified applicant has adopted and is using the trademark shown in the accompanying drawing for . . . .

(Name of Goods)

and requests that said mark be registered in the United States Patent and Trademark Office on the Principal Register established by the act of July 5, 1946, as amended thereafter.

The trademark was first used on the goods on . . . . was

(Date of First Usage)

first used in . . . . commerce on . . . . and is now in

(Type of Commerce)   (Date)

use in such commerce.

The mark is used by applying it to . . . . , and five specimens

(insert method of affixing.)

showing the mark as actually used are presented herewith.

```
State of . . . . . . . . . . )
                            ) ss:
County of  . . . . . . . )
```

. . . . , being sworn, states that he believes himself to be the owner

(Name of Applicant)

of the trademark sought to be registered; to the best of his knowledge and belief no other person, firm, corporation or association has the right to use said mark in commerce, either in the identical form or in such near resemblance thereto as to be likely, when applied to the goods of such other person, to cause confusion, or to cause mistake, or to deceive; and the facts set forth in this application are true.

```
                              Signed . . . . . . . . . . . . . . . .
                              (Signature of Applicant)


                              . . . . . . . . . . . . . . . . . . . .
       (Seal)                    (Notary Public)

  Date . . . . . . . . . . . . . . . . . . .
```

Fig. 15-2. An example of a trademark application by an individual (principal register with oath).

120

typed in capital letters or represented in the manner or configuration in which they are used on the goods.

Specification. The specification must be written in the English language and contains the applicant's name and address, the

Mark . . . .
(Identification of Mark)
Class No. . . . .

To the Commissioner of Patents and Trademarks:

. . . .
(Name of Applicant)

. . . .
(Full Address of Applicant)

. . . .
(Citizenship of Applicant)

The above-identified applicant has adopted and is using the trademark shown in the accompanying drawing for . . . . and requests that said mark be registered in the United States Patent and Trademark Office on the Supplemental Register established by the act of July 5, 1946, as amended thereafter.

(Type of Commerce)

. . . . commerce in connection with the goods for the year preceding the date of filing of this application.

State of . . . .)
) ss:
County of . . . .)

. . . . , being sworn, states that; he believes himself to be the
(Name of Applicant)

owner of the trademark sought to be registered; to the best of his knowledge and belief no other person, firm, corporation or association has the right to use said mark in commerce, either in the identical form or in such near resemblance thereto as to be likely, when applied to the goods of such other person, to cause confusion, or to cause mistake, or to deceive; and the facts set forth in the application are true.

Signed . . . .
(Signature of Applicant)

(Seal)

. . . .
(Notary Public)

Date . . . .

Fig. 15-3. An application to register on the supplemental register.

....
(Identification of Mark)

Class. No. ....

To the Commissioner of Patents and Trademarks:

....
(Corporate Name in State or County of Incorporation)

....
(Full Business Address)

....
(Situs of Corporation, Street, City, and State)

The above-identified applicant has adopted and is using the trademark shown in the accompanying drawing for .... and

(Name of Goods)

requests that said mark be registered in the United States Patent and Trademark Office on the Principal Register established by the act of July 5, 1946, as amended thereafter.

The trademark was first used on the goods on .... was first

(Date of First Usage)

used in .... commerce on .... and is now in

(Type of Commerce)

such commerce.

The mark is used by applying it to .... , and five specimens

(Method of Affixing)

showing the mark as actually used are presented herewith.

State of ....)

) ss:

County of ....)

.... , being sworn, states that he is .... of

(Name of Corporate Officer)   (Official Title)

applicant corporation and is authorized to execute this affidavit on behalf of said corporation; he believes said corporation to be the owner of the mark sought to be registered; to the best of his knowledge and belief no other person, firm, corporation or association has the right to use said mark in commerce, either in the identical form or in such near resemblance thereto as to be likely, when applied to the goods of such other person, to cause confusion, or to cause mistake, or to deceive; and the facts set forth in this application are ture.

....
(Corporate Name)

Signed ....
(Signature and Official Title)

(Seal)

....

Date ....   (Notary Public)

Fig. 15-4. An example of a trademark application by a corporation (principal register).

122

Mark . . . .
(Identification of Mark)
Class No. . . . .

To the Commissioner of Patents and Trademarks:

. . . .
(Name of Applicant)

. . . .
(Full Address of Applicant)

. . . .
(Citizenship, if applicable)

The above-identified applicant has adopted and is using the service mark shown on the accompanying drawing for . . . . and requests that

(Name of Service)

said mark be registered in the United States Patent and Trademark Office on the Principal Register established by the act of July 5, 1946, as amended thereafter.

The service mark was first used in connection with the services on . . . .
. . . . . ; was first used in the sale or advertising of services rendered in . . . .

(Type of Commerce)

. . . commerce on . . . . ; and is now in use in such commerce.
(Date)

The mark is used by . . . . and five specimens showing
(State method of using during service.)

the mark as actually used are presented herewith.

State of . . . .)
) ss:
County of . . . .)

. . . . , being sworn, states that he is . . . . of applicant's
(Name of Applicant)   (Title of Person)

service activity; he believes himself to be the owner of the trademark sought to be registered; to the best of his knowledge and belief no other person, firm, corporation or association has the right to use said mark in commerce, either in the identical form or in such near resemblance thereto as to be likely, when applied to the goods of such other person or services of such other person, to cause confusion or to cause mistake, or to deceive; and the facts set forth in this application are true.

Signed . . . .
(Signature of Applicant)

. . . .
(Notary Public)

(Seal)

Date . . . .

Fig. 15-5. An example of a service mark application (principal register).

11 to 13 inches long. The shorter side of the sheet is regarded as the top of the sheet and should contain the applicant's name, address, the date of first usage of the mark, and the types of goods on which it is used. The color of the drawing may be designated by matching it with the appropriate designations as shown in Fig. 15-1.

Fig. 15-6. An example of a trademark registration form.

UNITED STATES PATENT OFFICE

Scientific Enterprises, Inc., Los Angeles, Calif.

Act of March 19, 1920

Application October 15, 1945, Serial No. 489,989

ATOMIK

STATEMENT

To the Commissioner of Patents:

Scientific Enterprises, Inc., a corporation duly organized under the laws of the State of California and located at Los Angeles, California, and doing business at 5067 West Washington Boulevard, Los Angeles, California, has adopted and used the trade-mark shown in the accompanying drawing for a GAME EMBODYING A FOLDABLE GAME-BOARD MADE OF CARDBOARD AND HAVING ON IT DIAGRAMS OF VARIOUS ATOMIC STRUCTURES SHOWING THE ELECTRONIC CONFIGURATION IN ATOMS, AND IS PLAYED WITH MARBLES, PLASTIC DISCS, OR PAWNS, in class 22. Games, toys, and sporting goods, and presents herewith five specimens shown the trade-mark as actually used by applicant upon the goods and requests that the same be regis-

tered in the United States Patent Office in accordance with the act of March 19, 1920. The trade-mark has been continuously used and applied to said goods in applicant's business and in the business of applicant's predecessors since March, 1938. The mark has been in bona fide use for not less than one year in interstate commerce by the applicant or applicant's predecessors in business. The trade-mark is applied or affixed to the goods or the packages containing the same by printing said trade-marks thereon, or by placing thereon a printed label on which the trade-mark is shown

SCIENTIFIC ENTERPRISES, INC.,
By HRAND M. MUNCHERYAN,
President

Fig. 15-7. An example of a trademark.

The drawing should be transmitted flat to the patent office by using corrugated cardboard sheets on both sides of the drawing. Informal drawings may be accepted for examination, but new drawings must be made before the publication of the trademark. These are usually made by the patent office draftsman for a standard fee.

Specimens or Facsimiles. Five specimens of the mark as actually used on the goods should be furnished with the application. Facsimiles of the specimens may be furnished if the nature of the mark or its manner of affixing on the goods does not permit the transmittal of the specimens. Printed words as trademarks may be typed in capital letters or represented in the manner or configuration in which they are used on the goods.

Table 15-1. Classification of Goods and Services Under the Trademark Act.

Schedule of Classes: Goods

Class	Title
1	Raw or partly prepared materials.
2	Receptacles.
3	Baggage, animal equipments, portfolios, and pocketbooks.
4	Abrasives and polishing materials.
5	Adhesives.
6	Chemicals and chemical compositions.
7	Cordage.
8	Smoker's articles, not including tobacco products.
9	Explosives, firearms, equipments, and projectiles.
10	Fertilizers.
11	Inks and inking materials.
12	Construction materials.
13	Hardware and plumbing and steam-fitting supplies.
14	Metals and metal castings and forgings.
15	Oils and greases.
16	Protective and decorative coatings.
17	Tobacco products.
18	Medicines and pharmaceutical preparations.
19	Vehicles.
20	Linoleum and oiled cloth.
21	Electrical apparatus, machines, and supplies.
22	Games, toys, and sporting goods.
23	Cutlery, machinery, and tools, and parts thereof.
24	Laundry appliances and machines.
25	Locks and safes.
26	Measuring and scientific appliances.
27	Horological instruments.
28	Jewelry and precious metalware.
29	Brooms, brushes and dusters.
30	Crockery, earthenware, and porcelain.
31	Filters and refrigerators.
32	Furniture and upholstry.
33	Glassware.
34	Heating, lighting, and ventilating apparatus.
35	Belting, hose, machinery packing, and nonmetallic tires.
36	Musical instruments and supplies.
37	Paper stationery.
38	Prints and publications.
39	Clothing.
40	Fancy goods, furnishings, and notions.
41	Canes, parasols, and umbrellas.
42	Knitted, netted, and textile fabrics, and substitutions.
43	Thread and yarn.
44	Dental, medical, and surgical appliances.
45	Soft drinks and carbonated waters.
46	Foods and ingredients of foods.
47	Wines.
48	Malt beverages and liquors.
49	Distilled alcohol liquors.
50	Merchandise not otherwise classified.
51	Cosmetics and toilet preparations.
52	Detergents and soaps.

Schedule of Classes: Services

Class	Title
100	Miscellaneous.
101	Advertising and business.
102	Insurance and financial.
103	Construction and repair.
104	Communication.
105	Transportation and storage.
106	Material treatment.
107	Education and entertainment.

Specification. The specification must be written in the English language and contains the applicant's name and address, the date of adoption of the trademark and its use, the names of the particular products on which the mark is used, the mode of affixing the mark on the goods, and a separate oath or declaration for verification of the applicant's ownership, etc.

Filing Fee. A filing fee of $35 should accompany the application.

ACTION BY EXAMINER

An examiner reviews the application for a trademark to determine compliance with the statutory requirements and prior art. If the applications meets all requirements, and the prior-art search does not uncover a superseding trademark, the examiner allows the trademark registration. It will be issued a few months thereafter. The registration is good for 20 years and may be renewed thereafter within 6 months before its expiration.

In the event the examiner has required additional information to complete the examination, the applicant must respond within the time allocated by the examiner. If the applicant cannot meet the requirements within the allowed period, the application is declared abandoned. An abandoned application may be revived by a petition to the commissioner. A fee of $15 should accompany the petition.

INTERFERENCE AND OPPOSITION

As with a patent application, the examiner may declare an interference if two or more applicants are applying for the same trademark. The procedure is the same as in a patent interference. The examiner will instruct the applicant as to the next step to be taken if the applicant desires to pursue his case. If no opposition is filed within the time permitted for any interference action, the application will be granted and the trademark will be published in the *Official Gazette*. See Figs. 15-2 through 15-7 and Table 15-1.

Chapter 16
Copyright
Procedures

The copyright statute was enacted in the United States in 1790 for the purpose of fostering the public welfare by encouraging the creation and dissemination of intellectual works of authors and inventors. Another purpose of the original copyright statute was to grant an exclusive right to the creator of the intellectual work for making and selling his work by excluding others for a period of 14 years. By 1909, that copyright law had been amended three times. The law enacted in 1909 was in effect for 68 years thereafter (there were a number of unsuccessful attempts to revise the law). On September 22, 1976, a completely revised bill was passed, and on October 19, 1976, President Gerald Ford signed it into law. The copyright law, Public Law 94-553, became effective on January 1, 1978. The new law has shifted the direction of copyrighting through fundamental and pervasive changes. Many formerly conflicting problems that could not be resolved readily by the previous statutes were clarified. The new law establishes a single system of statutory copyright for published and unpublished works, and abrogates the earlier common-law copyright. The copyright protection subsists from the time the author's work is created.

WHAT IS A COPYRIGHT?

Under the present copyright law, in effect under the title 17 of the United States Code, a copyright grants protection to a person (having original authorship) the exclusive right to prevent others from copying the copyrighted work (whether published or unpublished). The work can consist of literary, musical, dramatic, or related publications. The law makes it illegal to violate the author's right by plagiarizing it in full or in part. Only the author or his authorized publishers can rightfully claim a copyright to a work. Joint authors of a copyrightable work are co-owners of the resulting copyright.

WHAT CAN BE COPYRIGHTED?

Any intellectual work that has become a tangible expression can be copyrighted under the present copyright statutes. This work may be either in writing, such as a book or a periodical, or reproducible by means of a machine or device such as a phonograph, motion-picture camera, or a television system. The following categories are included in the work of authorship as copyrightable:

☐ Literary works, such as books and periodicals. The application form to be requested from the Library of Congress is Form TX.

☐ Musical works, including any accompanying words. The application form for this is Form PA.

☐ Dramatic works, including any accompanying music. The application form is Form PA.

☐ Pantomimes and choreographic works. The application form is Form PA.

☐ Pictorial, graphical, and cultural works. The application form is Form VA.

☐ Motion picture and other audiovisual works. The application form is Form PA.

☐ Sound recording. The application form to be requested is Form SR.

An application form for any particular category in the above listed items can be obtained by writing to the Register of Copyrights, Library of Congress, Washington, DC 20559.

ITEMS THAT CANNOT BE COPYRIGHTED

There are several categories of materials that are not eligible for copyright protection because they fall into the public domain and are available to anyone without payment or authorization. Also, when the first 28 years have elapsed and no renewal has been made within the statutory period, anyone can use the material without regard to the originator or owner of the expired copyright. Upon expiration of a 28-year copyright, no copyright can be restored. Some of the examples of these categories are:

☐ Ideas, methods, systems, principles, methods of operation, concepts, discoveries, etc.

☐ Common or standard works such as calendars, measurement charts, lists of tables taken from public documents, and tables of chemical elements.

☐ Devices and blank forms such as those used for measuring and computing devices, slide rules, wheel dials, nomograms, mathematical principles, and equations. (Devices designed to record such expressions can be protected by patent laws.)

☐ Names, titles, slogans, variations of typographic ornamentations, lettering and the like. (These can be protected by trademark law under certain conditions.)

☐ Works of the United States Government produced by government employees. This includes material specifications, requirements for procurement by government agencies, and similar documentations.

COPYRIGHTING UNPUBLISHED WORK

In the old system, the protection of an unpublished work was afforded by common law. When the work was published, protection was provided by federal law. The present system is a single system of statutory copyright protection for all unpublished and published works. No notice of copyright is required on any unpublished work. Choreography and improvisations are still covered by common law until they are fixed in tangible forms and are subject to protection by federal statutes (if they are not published or registered). The advantages of the present system are as follows:

☐ Promotion of uniformity nationally. This avoids the difficulties of determining an author's right under different state laws. The enforcement of authors' rights is uniform.

☐ Outdates the legal significance of the word *publication*.

☐ Although the common-law protection of an unpublished work is perpetual, a time limit is placed on the duration of exclusive rights, Unpublished work becomes available to the public after a reasonable time (if the work is not copyrighted).

☐ The present system will improve the international dealings with respect to copyrights.

☐ Any unpublished work that is existent since January 1, 1978, and has not been in public domain nor has been protected by a statutory copyright, is automatically protected by the new statutory system. The work enjoys the same rights as if protected by a statutory copyright (within a reasonable time period).

The procedure for copyrighting an unpublished work consists of depositing one complete copy of the unpublished manuscript in the

Library of Congress, together with a fee of $10. When the work is published, two copies of the published work should be deposited in the Copyright Office, in addition to the unpublished manuscript already deposited.

COPYRIGHTING PUBLISHED WORK

When the work is published (distributed or sold generally), two copies of the published work must be deposited with the required fee in the Library of Congress. The proper application should be selected and filled out completely to include the date of first publication and whether the publication has occurred in the United States or in a foreign land that is a member of the Universal Copyright Convention. The owner of such a copyrighted or the party authorized by the copyright owner can reproduce, prepare derivatives, distribute, rent or lease the copyrighted work to any party he deems desirable. It must be pointed out here that any unauthorized use of the copyrighted work for purposes such as criticism, comments, news reporting, teaching, or research work does not constitute infringement under the new copyright statutes. Nevertheless, such use must be exclusively for nonprofit, educational purposes. All published works must have the copyright notice properly affixed.

DURATION OF COPYRIGHT

The duration of a copyright secured after January 1, 1978 is for the life of the author plus 50 years after his death. In the case of joint authors, the 50-year period is measured from the date of the death of the last surviving author. The works of anonymous authors may bear a copyright for a term of 75 years from the date of the first publication or 100 years from the date of its creation. This stipulation holds for any work made for hire. Work prepared for hire signifies that an employee has prepared the work during and within the scope of his employment by an employer.

In the case of an unpublished work in existence on January 1, 1978, not protected by statutory copyright and not in public domain, the present copyright law provides automatic federal protection for the life of the author plus 50 years. All works in this category are provided at least 25 years of statutory protection. If the work is published before the expiration of the 25 years, the term is extended by another 25 years.

Any work that has been copyrighted before January 1, 1978, and where the copyright is still subsisting, the protection of the copyright continues to the end of the first 28 years (as per old law), and the work is renewable for another 47 years, extending the total term of the copyright to 75 years. The renewal application must be made within one year before the expiration of the first 28 years of copyright. This provision was included to avoid confusion in an existing author-publisher contract and the publication of authors' work.

COPYRIGHT NOTICE

The new copyright law, similar to the old law, requires that a copyright notice be affixed on every copy of the work that is published in order to notify the public that the work is protected by a copyright. With the old law, if the copyright notice was left out inadvertently or purposely, the owner of the copyright would permanently lose his right for a copyright. The new law has abolished this penalty and provides that, if the omission of the copyright notice is inadvertent, subsequent copies of the published work can be marked without loss of copyright protection.

The copyright notice consists of the word "Copyright," the year of the copyright, and the name of the copyright owner. The notice should appear on the work even when the work is published outside of the United States. The marking of the copyright on the published work is the responsibility of the copyright owner. Marking the word "copyright" on the published work before publication of the work does not require permission from the Copyright Office.

The location of the copyright notice should be such that it would be readily noticeable by any person using the copyright article. On a book, the notice usually appears on the reverse side of the title page. On sheet music, it appears on the first page. On a phonographic record, it appears on the label affixed at the center of the record. On periodicals, it appears on the content page. On other printed matter, it should be located on a page and an area easily noticeable such as a section of the first page or its reverse side where the publisher's name is given. Because the instructions of a game are subject to a copyright, the notice should be given adjacent the manufacturer's name and address.

The copyright notice consists of three parts: the word "Copyright," the year, and the name of the copyright owner. For example, Harry Smith has copyrighted his book on January 4, 1980. The notice will read: Copyright 1980, by Harry Smith. When space is of concern, the copyright notice may also be written as follows: a capital C enclosed in a circle, followed by the year of copyright, and the name of the copyright owner. For example, © 1980, Harry Smith, in case the publication is a book, periodical, game, one-sheet brochure, a printed speech and the like. For sound recordings, such as a phonograph record, the capital "P" encircled is used. For example, ℗ 1980, Harry Smith.

COPYRIGHT OFFICE FEES

An application form obtained from the register of copyrights should be completely filled out, signed, and mailed (with the required number of copies of the published work) together with the applicable fee to the Register of Copyrights. The various fees are as follows.

☐ For registration of a newly published book: $10.

☐ For renewal of a book copyright: $6.

☐ For registration of transfer of copyright ownership, consisting of one page: $10. For additional sheets in excess of one: 50 cents for each sheet.

☐ Issuance of any certification on copyrighted matter: $4.

☐ For making any Library search with regard to any prior copyrights: $10.00 for each hour and fraction thereof.

REGISTRATION PROCEDURE

Registration in the copyright office is for the benefit of the owner of the literary or related work. It establishes public record of an author's copyright claim, and it is prima facie evidence of ownership of the work. When an infringement proceeding is to be taken, the work without registration will nullify the author's claim. Registration may be made within three months before the infringement suit is instituted. The author should not wait to register until an infringement becomes necessary. The copyright registering procedure to pursue is as follows:

☐ Complete the application that is applicable to the work in hand.

☐ If the work is unpublished, submit one copy with the application and the fee.

☐ If the work is published in the United States on or before January 1, 1978, two complete copies should be provided with the proper fee.

☐ If the work was first published outside of the United States, one complete copy should be submitted when the work is published in the United States, accompanied by the proper fee.

□ The copyrightable work, the application, and the fee should be included in one package for the submittal to the Library of Congress.

COPYRIGHT INFRINGEMENT: GENERAL

Anyone violating the exclusive right of a copyright owner is an *infringer* and is subject to an *infringement suit.* Damages and heavy penalties may be imposed under the present copyright laws. The law grants the courts the power to impound allegedly infringing materials during the pending court action. The court is also empowered to order the destruction of the infringing articles. An infringer of a copyrighted article is subject to defray the sales damages incurred by the copyright owner, and reimburse the copyright owner all the profits and income derived by the infringer as a result of his making and selling the infringed articles. In addition, the court may impose penalties as high as $50,000. This depends on whether the infringement committed arose from fraudulant and false representation. The new copyright law, in general, is more severe to the infringer than the law embodying the infringement of a patent. The statute of limitation for bringing a legal suit for the infringement of a copyright is three years.

INFRINGEMENT AS A CRIMINAL OFFENSE

The present copyright law states that any person found guilty of infringement committed "willfully and for the purpose of commercial advantage or private financial gain shall be fined not more than $10,000 or imprisoned not more than one year, or both." In the case of infringing sound recordings, the infringer "shall be fined not more than $25,000 or imprisoned for not more than one year, or both, for the first offense, and shall be fined not more than $50,000 or imprisoned for not more than two years, or both, for any subsequent offense." When the infringer is convicted of any violation of the preceding conditions of copyright protection of a copyright owner, the court may also order the forfeiture and destruction of all the infringing copies of the copyrighted articles.

MANUFACTURE OF COPYRIGHTED WORK

Prior to July 1, 1982, any importation of any article copyrighted in the United States and manufactured abroad was prohibited, and no copyright protection was afforded to foreign authors. For example, the work of English authors could be pirated by imprudent publishers in the United States. The new copyright act permits the copyright owner in the United States to have his work printed abroad, photocopies made there, and to import the photocopies from which printing plates may be made in the United States for the publication of the author's work. The foreign authors have the same right as the domestic authors to have their work copyrighted in the United States and be afforded the same protection as for authors domiciled in the United States.

ASSIGNMENT OF A COPYRIGHT

A copyright can be assigned by the owner to a second or third party at some remuneration agreed upon by the parties entering into the assignment contract. The new law has a number of stipulations that may be made a part of the contract in the assignment. The parties entering into such a contractual binding must write to the Register of Copyrights requesting a copy of the Chapter 8, Copyright Royalty Tribunal, Sections 801 through 810, for study and abstraction of the part related to the contractual stipulations.

Section 602 of the Tribunal identifies unauthorized importation as an act of infringement, and permits the Custom Service to prohibit importation of only "piratical" articles (copies of the copyright owner's work made without any authorization of the owner). In other words, under this section, unauthorized importation of the copyright owner's work constitutes infringement. It is commendable that the present copyright law not only is severe, but also makes the infringement a concern of the federal responsibility and action. This is unlike the present law covering patent protection. There the sole responsibility belongs to the patent owner.

Chapter 17

Filing for a Patent in a Foreign Country

A United States patent right extends only throughout the United States and its territories. An inventor who wants to patent his invention in any of the 79 foreign countries that are members of the Paris Convention for the Protection of Industrial Property must file a separate application in each of the nations selected. This treaty extends the same patent rights for the patentee as when he obtained a United States patent. The treaty further provides the right of priority as in cases prosecuted in the United States. This means that when an applicant files a patent application in any one of the member countries, his date of filing will apply in the consideration of priority. The filing date will be regarded as the date of application in any future application in another country. A foreign application for a patent must be filed within 12 months after an application is filed in the United States Patent and Trademark Office. In the case of design patents and trademarks, the period for filing is six months.

A more recent treaty signed by 35 countries (in Washington, D.C., in 1970, and became effective on January 24, 1978) is known as the Patent Cooperation Treaty. This treaty facilitates the filing of a patent application on the same invention and provides a centralized filing procedure and standardized application formats. The inventor must contact someone in the foreign country who is familiar with filing of patent applications. Patent offices in some countries do not have the facilities to correspond with the inventor directly.

FOREIGN LICENSES

An inventor who wants to file for a patent in a foreign country before he applies for a patent in the United States must first obtain a license from the Commissioner of Patents and Trademarks. If a patent application has already been filed in the United States Patent and Trademark Office, the license must be obtained within six months after the date of filing. An inventor may write a simple letter stating the facts regarding his invention, or he may identify the application already filed by reference to the serial number, date of filing, and the applicant's name. If the application has been filed in the United States patent office for more than six months, no license is necessary for foreign filing. On the other hand, if the patent application is being detained by the patent office due to a secrecy order, no patent application can be filed in a foreign country until the secrecy order is lifted.

PENALTY FOR FILING WITHOUT A LICENSE

Within the statutory period of six months, any person who files an application for a utility or design patent in a foreign country without securing a license from the Commissioner of Patents and Trademarks is barred from obtaining a patent in the United States. If the patent has already been obtained in the United States without previous knowledge of the patent office, respecting the foreign filing, the patent thus issued in the United States is considered invalid. Whoever—during the entire period of secrecy order declared on his patent application, and with knowledge of such order—willfully publishes or discloses, or causes to be disclosed, or files any patent application on the invention in a foreign country, will be fined, when convicted, not more than $10,000 or imprisoned for not more than two years, or both. This penalty does not apply, however, to any officer or agent of the United States Government acting within the scope of his authority, or acting upon authorization by written permission by the United States government.

FILING PROCEDURE IN A FOREIGN COUNTRY

It will be of benefit to the inventor to consult an attorney in the United States for filing a patent application in a foreign country because the laws and regulations in different countries vary and that the attorney may have an associate or agent located in the foreign country to attend to the prosecution of the application. Some countries do require that the application be translated to the language of that country before filing. This procedure is often within the capability of a patent attorney who is specialized in foreign filing. Either the attorney or a person he knows would be familiar with the translation, and especially the claims that require the proper scope and spirit as found in the application filed in the United States. In translation, if the claims are inadvertently altered in their scope, the strength of the patent that is issued may be diminished compared with the original claims. For this reason, a competent attorney's services should be obtained.

FOREIGN APPLICANTS APPLYING FOR U.S. PATENTS

Thousands of foreign inventors apply for United States patents annually. The United States patent laws make no distinction with respect to the country where the inventor is a citizen. Any inventor from a foreign country can apply for a patent on the same basis as an American citizen.

As for all United States citizens, the application papers must be signed by the inventor, as described in this book, even though such signatures might not be necessary in the country where the inventor is domiciled. All rules and regulations that apply to a United States citizen also apply to foreign inventors applying for patents in the United States. No United States patent may be obtained if the invention has been patented in a foreign country 12 months prior to the filing of a patent in the United States Patent and Trademark Office. For design patents, this rule applies for six months prior to the application for a design patent in the United States.

Quoting from the General Information Concerning Patents in the United States, the document states: "An application for patent filed in the United States by any person who has previously regularly filed an application for a patent for the same invention in a foreign country which affords similar privileges to the citizens of the United States shall have the same force and effect for the purpose of overcoming intervening acts of others as if filed in the United States as on the date on which the application for patent on the same invention was first filed in such foreign country, provided the application in the United States is filed within 12 months (6 months in case of design patents) from the earlier date on which any such foreign application was filed. A copy of the foreign application certified by the patent office of the country in which it was filed is required to secure this right of priority."

An oath or declaration must accompany the application. When an oath is made, in accordance with the sample oath given in Chapter 4, the applicant must state the country in which the earliest such application has been filed and give the filing date of the application. This statement should be added to the end of the oath before the signature of the applicant. The oath should be signed and dated by a notary public or an official of the government who is entitled to notarize papers in the foreign country. If a declaration is made (See Chapter 4), no signature of a notary or any government official is necessary.

When an oath is taken before a notary public or an officer authorized to sign formal papers, the entire application papers (except the drawing) must be stapled together, at least three holes must be punched near the staples to pass a ribbon through all of them at least twice and they must be sealed officially. Every sheet in the application should bear the official seal of the authorized person attaching the papers. When the applicant uses a declaration form, (see Chapter 4), the ribboning procedure may be dispensed with and no official signature or seal is necessary.

For the convenience of the inventor, after he has prepared his application in accordance with the instructions given in this book, he may hire a patent attorney (in his country or preferably in the United States) and have the attorney handle the filing. This procedure will save hundreds of dollars for the inventor. The patent attorney will charge a small fee for the time he spends for filing and corresponding with the inventor. The patent office will not correspond with the inventor in a foreign country when the case is filed by the attorney. An attorney in the United States will have additional advantages because, at times, the examiner will allow only 30 days for the applicant to respond to his office action. Those 30 days, at times, may be used up for transmittal of papers to and from the patent office. Such a situation will not leave sufficient time for the inventor to prepare his response. If a United States patent attorney is needed, the inventor should write beforehand to the United States Patent and Trademark Office, Washington, D.C. 20231, U.S.A. and ask for a copy of the roster of registered patent attorneys. A fee of $3.70 in United States currency should accompany the request.

Chapter 18
Drafting
Patent Drawings

In almost all patent cases, the applicant is required to furnish a drawing of his invention. The drawing is submitted with the patent application. In cases of chemical compositions, a drawing might not be necessary unless a graphical representation of a new organic compound has to be made to show attachment of the side radicals. A drawing might also be necessary for a chemical processing machine, flow lines, or steps that are to be shown in describing the process. The drawing must show every feature of the invention claimed in the application. The drawings facilitate the understanding of the invention by the examiner as well as by the general public when the application becomes a patent. The drawing should be made at a scale large enough to show the mechanisms without crowding.

CHARACTERISTICS OF THE DRAWING

Type of Paper Used. The patent drawings should be made on white, two-ply or three-ply Bristol board. The surface of the paper must have been calendered so that erasures can be made. India ink must be used to make all drawings. A razor blade should not be used to remove the india ink because this usually leaves a rough surface and the ink might spread out and make the lines fuzzy. The use of white pigment to erase errors is not acceptable by the patent statutes.

Size of Drawing Paper. For utility patent drawings, the size of the drawing paper should be 8½ by 14 inches (21.6 by 35.6 centimeters). A top margin of 2 inches and bottom and side margins

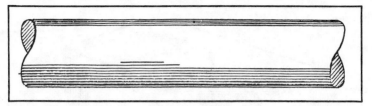

Fig. 18-1. Surface shading for pipes and shafts.

of ¼ inch each must be included. The drawing should be made within the 8-by-11¾-inch area. No marginal lines should be drawn. For trademarks, the drawing sheet must be 11 by 13 inches and should include a top margin of 1 inch, a left margin of 1 inch, a right margin of ¾ of an inch, and a bottom margin of at least ⅜ of an inch.

Drawing Views. The drawing must contain as many views of the invention as necessary to show its general construction and appearance. Each individual view of the drawing should be labeled with a figure number. The figures may be a perspective view, plan view, cross-sectional view, or a detail view of a small section drawn on a larger scale. Exploded views may also be drawn and embraced by a bracket to show the interrelationship of the separated elements. When necessary, a large machine may be drawn on several sheets. The same is true with electronic circuit diagrams. These may be broken at identifiable portions and drawn on separate sheets. Each sheet would be consecutively numbered and the input and output signal points marked identifiably. The views of the drawings must be described by a short paragraph in the specification.

The various sections of each view should be numbered consecutively. An identical part appearing in more than one figure should bear the same reference number. This is true unless it is an analogous part that is slightly modified. In such an event, a new reference number should be assigned. Parts in a drawing or in a view that are common knowledge can be represented by a block and numbered. If an unconventional drawing of a part is made, it should

Fig. 18-2. Shading for spherical objects.

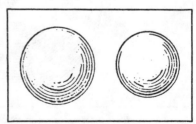

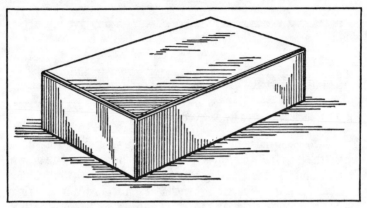

Fig. 18-3. A rectangular block in perspective.

be identified by a word or words. This practice is not desirable if too many such markings are on the drawing.

Hatching and Shading. Hatching should be made by using oblique, parallel lines not less than 1/20 of an inch. Heavy and light lines may be drawn for shading in accordance with the examples in the Examples for Drafting section of this chapter. All hatching and shading should be done using drafting instruments. Freehand lines or drawings should be used only when other methods are not available. In shading a part of the drawing, the shading should conform to a delineation assumed to be illuminated from a left-hand corner at about 45 degrees.

Reference Characters. The reference characters (labels) should be leroyed and numbered consecutively. Each character

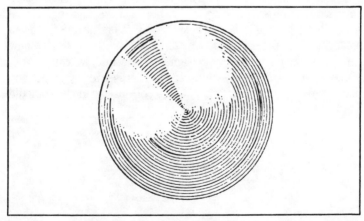

Fig. 18-4. Surface shading for a conical object.

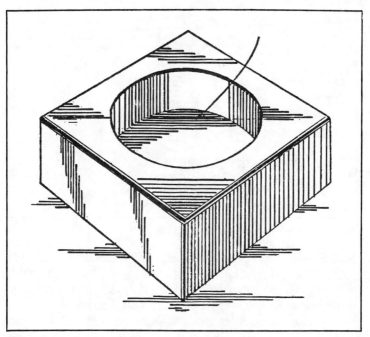

Fig. 18-5. Shading for an ellipse on a horizontal surface.

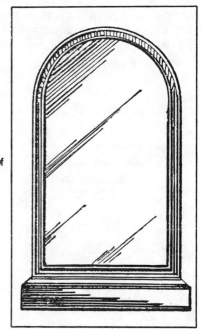

Fig. 18-6. Shading on the surface of a mirror.

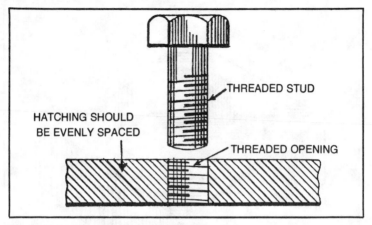

Fig. 18-7. An example of illustrating threads.

should be at least ⅛ of an inch in height so that when the drawing is reduced for final printing it will not be less than 1/24 of an inch. The numbers should be placed legibly without crowding to avoid misunderstanding. All lines should be clean, sharp, and solid.

No reference character should be placed on shaded or hatched surfaces. It is preferred that each reference number be separated from the other clearly and delineated by a line pointing to the part being described (without using an arrowhead). A whole part containing a number of reference numerals may be pointed out with a line having an arrowhead. The reference characters may be either numerals or capital letters, but capital letters should be kept to a minimum because the number of parts to be designated might be beyond 25 and there are not enough letters to accommodate such numbering. See Chapter 5 for examples of numbering.

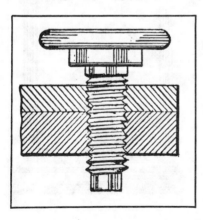

Fig. 18-8. Shading for round handles and screw threads.

144

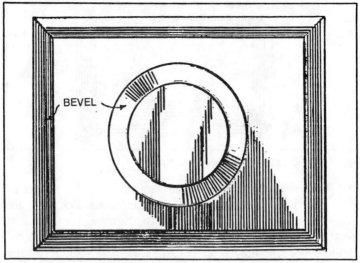

Fig. 18-9. Shading on bevel edges.

AMENDMENT TO THE DRAWING

After the drawing is filed with the application in the patent office, no changes in the drawing may be made except by requirement of the examiner for clarification. A sketch made with perma-

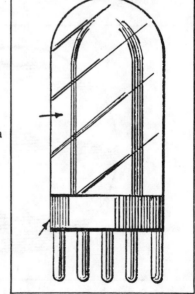

Fig. 18-10. Surface shading on a radio tube.

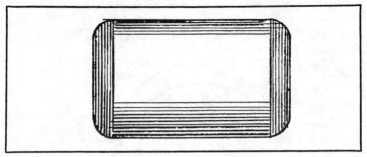

Fig. 18-11. Highlights of cylindrical shading.

nent ink showing the proposed changes must be filed in the patent office and, if in the opinion of the examiner such changes are acceptable, the office draftsman will make the change or addition at the standard service charge. Substitute drawings usually are not admitted because they require extra time and scrutiny by the examiner and errors might enter inadvertently. Any drawing informality or errors by the applicant are objected to by the examiner. The applicant will be notified for correction on a separate sheet for submittal. For this purpose, the applicant should make several extra photocopies of his drawing before he files it with the application. Any changes can be made with a red pen on the copy of the drawing and submitted to the examiner for his approval and entered in the official drawing.

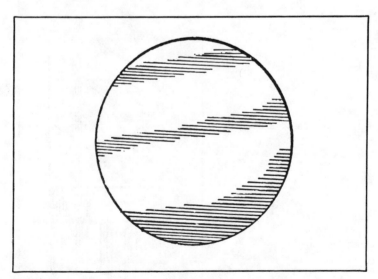

Fig. 18-12. Surface shading for a discal plate.

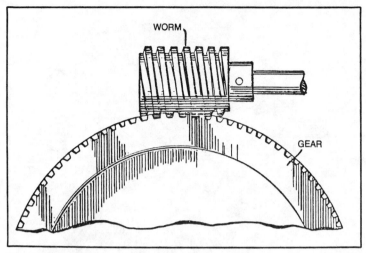

Fig. 18-13. Illustrating a worm gear aspect.

DRAFTING SYMBOLS

Conventional graphic symbols may be used on the drawings. Electronic circuits may follow the standard electrical and electronic symbol formats. Mechanical symbols, when not within the conventionally accepted standard forms, may be objected by the examiner—who will instruct the applicant accordingly. Such symbols may be altered by the patent office draftsman upon a request by the applicant and accompanied by the required fee.

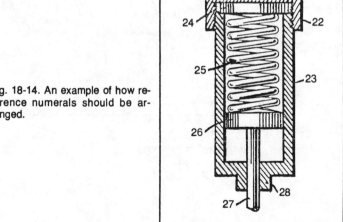

Fig. 18-14. An example of how reference numerals should be arranged.

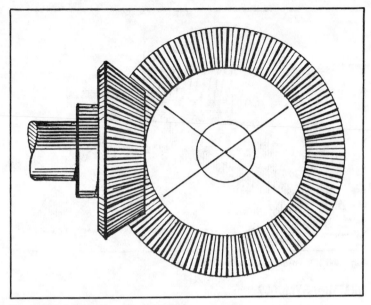

Fig. 18-15. A plan view of a bevel gear.

EXAMPLES FOR DRAFTING

A series of standard drafting forms are included in this section. These forms are presented along the formats of the patent office draftsman's guide. The inventor should use the formats as closely as possible to delineate the structure and shading.

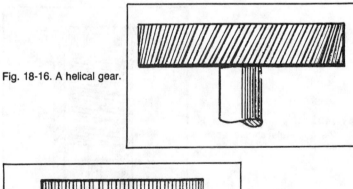

Fig. 18-16. A helical gear.

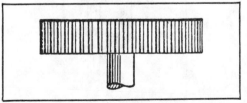

Fig. 18-17. a spur gear.

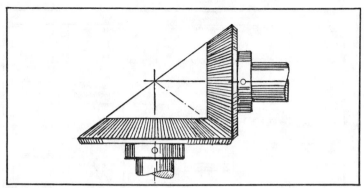

Fig. 18-18. Bevel gears with similar slant gears.

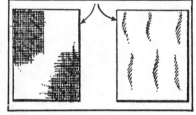

Fig. 18-19. Two methods of illustrating fabric.

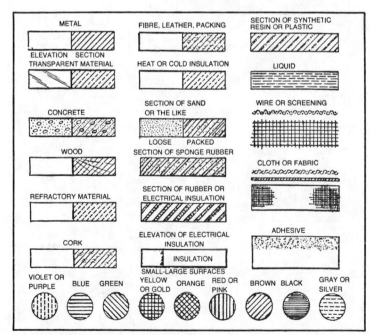

Fig. 18-20. Various symbols to guide draftsmen.

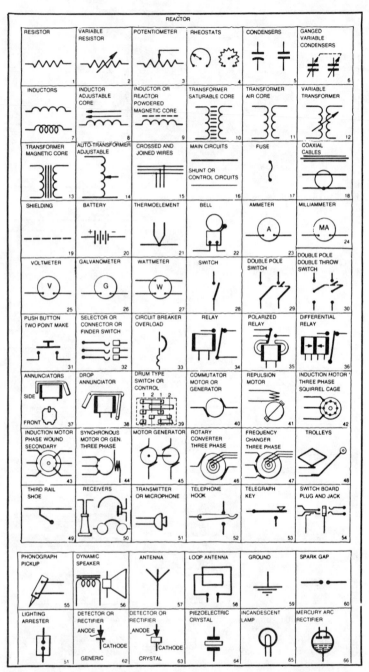

Fig. 18-21. Electrical symbols.

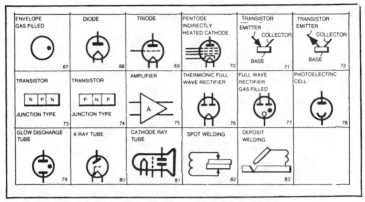

Fig. 18-21. Electrical symbols. (Continued from page 150.)

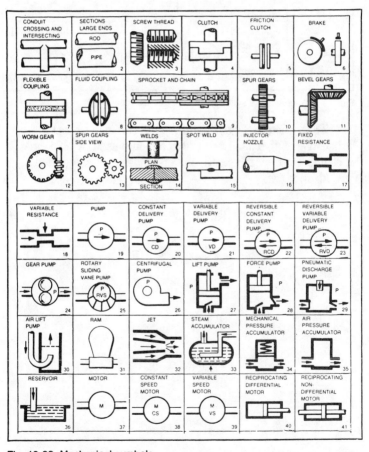

Fig. 18-22. Mechanical symbols.

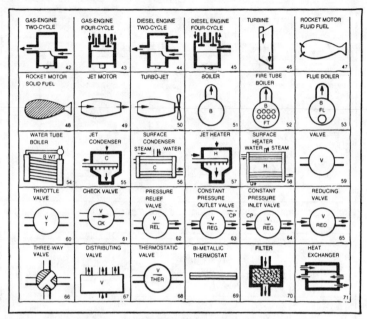

Fig. 18-22. Mechanical symbols. (Continued from page 151.)

The symbols (Figs. 18-1 through 18-22) in this chapter are not exhaustive. Other standard and commonly used symbols may be acceptable if they are clearly understandable and identified in the specification.

Chapter 19
Employer/Employee Rights to an Invention

Who owns the invention when a person invents something during the course of his employment by a company? The consensus of opinion entertained by the general public is that the ownership in the invention is vested in the employer. Surprisingly, the fact remains that the employee owns the rights to an invention whether or not he invented it during his employment by the company and with the aid of company's facilities and materials. The inventor is not obligated to assign his invention to the company unless there is a specific stipulation in his employment contract that calls for the assignment of any product he invents at work whether the invention is related to company-manufactured products or not. Even then, the second part of the last statement might not hold in courts when contested.

When a person is employed for the express purpose of inventing and assigning his inventions to the employer and such is the essence of an agreement made by and between the employer and the employee, then the employer has every right to own the invention. If patented by the employee during his employment, the patent rights are assignable to the employer. The assignment of the patent should be performed by writing, and submitting a copy to the patent office and recording it in the name of the employer as the assignee. In the absence of any agreement specifying the provisions expected of the employee arising from his employment obligations, the employee can refuse to assign any invention, patent application, or a patent to the employer. A blanket contract while securing the employment by the employee stating that any product invented by the employee in the course of his employment belongs to the employer would be against public interest. Under such conditions, the employee would have no incentive to invent and the progress in inventive science would be retarded. In this respect, some companies have in their company policies a provision to remunerate the inventor in their employ with a bonus or an advance in salary. Such an arrangement will encourage the employee to think, design, and develop inventions and then assign the patents, when granted, to the company. In addition to his compensation, it will be an added satisfaction to the inventor to see his invention produced and sold by his company.

In the absence of a distinct understanding between the employer and the employee in writing and a definition of mutual obligation—prior to the entrance of the employee into the service of the employer—any invention made by the employee on his own time or on company time using company tools and instruments legally belongs to the employee. This rule includes the officers of the company and even the president or the general manager. If the invention or the improvement is related to company business, then the inventor is obligated to assign any right to the company to make, use, and sell the product without any special compensation to the inventor (except his regular salary). The company could compensate the inventor by some other means at its own discretion.

All rights of the invention made by the employee outside of his employment hours and not related to company business are automatically vested in the employee without regard to his employer. On the other hand, an inventor might want to approach his company with his idea or invention in an attempt to interest his company in purchasing the invention or its patent rights. If no patent or application for a patent exists, the company might want to undertake the patenting of the invention. In such an event, the inventor should sign and submit a formal assignment and record it with the patent office. The submission to record the patent office should be made within three months after signing of the assignment. Otherwise, the assignment will be difficult to sustain even in the courts when contested by the buyer. The law states: "An assignment, grant or conveyance shall be void as against any subsequent purchaser or mortgagee for a valuable consideration, without notice, unless it is recorded in the Patent and Trademark Office within three months from its date or prior to the date of such subsequent purchase or mortgage." (USC Title 35, Patents, Section 261.)

The assignment should identify the patent by number, date of issue, name of the inventor, and the title of the invention. When the assignment is for an application, it should be identified by serial number and date of application, and have the name of the inventor and title of the invention. A recording fee of $20 should accompany the assignment.

SHOP RIGHT BY EMPLOYER

In the absence of a contract, if the invention is applicable to the employer's products of manufacture and is produced in the employer's facility using the materials and tools of the employer, then the employer will be entitled to a *shop right*. This means that the employer has an unwritten license from the employee and has the right to make, use, and sell the product without compensation to the inventor. If a patent is granted on the invention, then the patent belongs to the employer's business. When the business is sold, the right to the invention is transferred with it. However, the business cannot give out license to other manufacturers on the patented product. No formal execution of documents is necessary. Nevertheless, the right of the employer or his successor should be documented and recorded at the Patent and Trademark Office in the name of the owner.

JOINT OWNERSHIP OF A PATENT

A patent may be owned jointly by two or more persons. In such a case, either both of the owners have invented the invention or one of the owners has invented and the other helped him financially to construct the invention or defrayed the cost of patenting. In doing so he has been assigned a part interest in the invention. Each joint owner of the patent has the right to make, use, and sell or grant licenses to others without the consent of the co-owner unless a contract has been written binding both co-owners of the patent to certain conditions or stipulations. It is not advisable to assign a part interest in a patent to another person without a definite agreement delineating the rights of each of the co-owners. If the invention or the patent is to be held jointly by the joint owners, so that any remuneration derived from it will be shared equally, then a clause should be inserted in the agreement to the effect that the interests of both parties are undivided, and that no one party will make, use, or sell the invention without sharing the profits equally with the other. In such an event, it would be best for the parties involved to hire the services of a competent patent attorney familiar with patent contracts.

If an inventor develops his invention in his employer's plant under the supervision of his department head or supervisor, such persons cannot be considered as joint inventors unless the supervisors contributed to the invention of the product. The inventor then must sign all the patent papers and the application as a sole inventor. If the inventor and a supervisor sign a patent application as joint inventors and later it is found that the supervisor is not a joint inventor, the patent may lapse.

EMPLOYMENT CONTRACTS

Many engineers and scientists employed by large corporations work for the express purpose of inventing, designing, and developing products that fall within the scope of the company's activity. Such persons usually must sign an agreement that any item that they invent must be assigned to the company. This type of arrangement is perfectly fair and legal. In the long run, it might even help the inventor move up to a higher level of employment. The inventor has to sign all documents related to the securing of patents on the products he invents, develops, or improves. In rare occasions, some companies make awards of regular sums to the inventor each time he has produced a new development that the company can commercialize.

While there are some advantages to this arrangement there are also disadvantages because the employee-inventor will have to work under pressure to meet scheduled programs and his products are expected to be commercially successful. Furthermore, the employee is directed not to disclose, use, or patent any product during or subsequent to his employment. The last condition of employment poses a new conflict to the life of the employee-inventor. Such stipulations are generally known as *trailing clauses* in the employment contract. These should not be signed without qualification

Trailing Clauses

Perpetual agreement to assign all inventions to a former employer is against public policy and will not hold in the courts when contested by the employer. No person is obligated to perform any work for a former employer after severance of employment and without compensation for his efforts. Some companies state in their employment contract that the person, subsequent to leaving the company permanently, will not for at least six months, invent or patent any invention related to the company business or that has any bearing on the knowledge obtained from the company's business. Such a stipulation is a useless expedient. If the former employee has conceived a new invention and has applied for a patent, there is no way of finding out about it. The patent office will keep the matter of patent application secret, and no information would be given out without the inventor's signed consent.

Agreement to Only Company-Related Inventions

Recently many companies are turning toward a more equitable arrangement with the prospective employee-inventor. The employee is instructed during his employment interview that he is being hired to advance the employer's technology and that he is expected to assign to his employer any and all new inventions related to the employer's business that he conceives during his employment. Such an agreement is equitable and not unreasonable to require of the new employee. The employer is anxious to have his product improved or have new products developed related to his line of manufacture to keep the company's competitive position against other companies producing the same line. Any patent that is granted on such inventions under the conditions of the employee's employment must be signed by the employee-inventor. If he refuses, the courts cannot force him to sign. Therefore, the condition of signing all documents should be specified in the employment contract. The employee should be given a copy of the contract.

It would be preferable for the employer and the employee to prepare a patent application prior to assignment of the rights to the employer. The advantage of such an arrangement is that the invention can be definitely identified by name, genus, date of conception, serial number, and the inventor's identity. In this manner, an analogous invention conceived later by the employee-inventor can be differentiated from that for which a patent application has been filed. If the employee has already severed his employment, then whether the second invention should be assigned to the former employer or the inventor can only be determined by the employment contract stipulation.

PATENT APPLICATION SIGNED BY INVENTOR ONLY

All applications on inventions that are to be patented must be signed by the true inventor, even when the invention has been assigned to the employer either verbally or by a written instrument.

If the inventor has left the company before the application was filed, he still is obligated to sign the patent application. The patent will be issued to his name with his former company as the assignee. If the inventor cannot be located, a company employee who was closely associated with the invention and had helped in the development of the invention can sign the application as joint inventor. In such an event, a separate document should be directed, together with the application, to the Commissioner of Patents and Trademarks. State the fact that the other inventor cannot be located.

SCIENTIFIC INVENTIONS CONCEIVED BY ACADEMICIANS

Instructors or professors working in the research laboratories of colleges or universities at times conceive patentable inventions in the course of their research. These may be commercially useful products. For example, osmosis is a law of nature that is not patentable, but discovery of reverse osmosis and the means for producing it would be an invention of the first magnitude (if no prior art exists). The invention would then be subject to patentability. Who pays for the patent application and who owns the patent rights if the item is patented?

Some universities do not allow their employees to patent an invention for profit. Others pay for the expenditure and compensate their employee-inventor with a lump sum or by giving the inventor a royalty if the invention is commercialized. Some institutions have been successful in exploiting the inventions developed by their employees and compensating them amply by issuing stock or cash advance. In these instances, the inventions made by the employees during the course of their employment and using the facilities of the university are assigned to the university's nonprofit subsidiary. Both the university and the employee-inventor benefit. Because policies differ with respect to invention compensation from one university or college to another, it would be best to find out about the institution's attitude toward inventions and patents before becoming involved with inventing and developing on university time and expenditures.

CARE IN DRAFTING CONTRACTS

Great care should be taken in drafting a contract for the sale or licensing of an invention. If the buyer or licensee is an employer and the contract is too broad, such as, "all rights of any inventions made by the employee will be vested in the employer (a phrase that does not specify the types of inventions, their relationship to the employer's business, the time or period during when the invention was conceived and developed, or under whose sponsorship), the contract will be declared void if adjudicated. If the contract stipulations are too restricted, the contract does not cover all the related inventions developed by the employee during his employment. When the terms of the contract are not clear, then the employee owns the invention even when the knowledge to conceive the invention was derived during his gainful employment.

As an example, a chemical professor worked for a universitv with a subsidiary nonprofit manufacturing corporation that handled the patented new chemical mixtures or compounds developed by the university professors or employees. The employees were under contract to the university. The contract in part stated that "any future chemical formulas, compounds, chemicals, and mixtures, or inventions related to dyes, colloids, paints, and any improvements thereof," will be assigned to the university. Subsequent to the signing of the contract, one instructor invented an electronic transducer for measuring flow of liquids through conduits. The university claimed that the invention belonged to the university because it was conceived during the inventor's employment at the laboratory and the inventor used the laboratory facility to develop the item on university time. The case was taken to court. The inventor was allowed to keep his invention because his contract read only on developing chemical compounds and improvements thereof.

Such experiences with employee-employer relations show that any contract between the employer and the employee should be written in clear, definite, and complete terms to avoid any controversies that arise in the future as to who owns the rights of the new improvements and the inventions developed during one's employment at a company or institution. Were a phrase such as ". . . and other mechanical, electromechanical, and electronic invention" included in the employee's signed contract, then the university would have had every right to declare the invention as its own.

With regard to the outcome of the preceding court case, it must be kept in mind that while the case may establish a precedent for similar court cases in the future, a good attorney can reverse that precedent on the grounds that the invented transducer was a part of the employee's responsibility in carrying out his duty with relation to his development of chemical compounds or mixtures. For instance, while working with the chemicals and transmitting them from one vessel to another, a commercial gauge was used to measure the flow rate of the chemical. The employee of the institution improved on the gauge and produced a more sensitive and accurate device. The improvement on the existing products was the employee's responsibility in order to conduct his duties efficiently and effectively. Therefore, the flow-measuring device was a result of his advancing the art of developing chemical products. Accordingly, the new so-called invention still could belong to the university. It goes to show that each case should be considered on its own merits rather than being dependent on a precedent.

GOVERNMENT EMPLOYEE PATENT RIGHTS

Any military enlisted man, officer, or any government employee, except the officers and employees of the Patent and Trademark Office, may be granted a patent on an invention without payment of patent fees. The right of the patent issued to the inventor is vested in the United States government, and is used by it without payment of royalty to the inventor. The government, however, may waive certain rights to the inventor to license the invention or the patent to outside manufacturers if the invention is not classified. The law, 42 USC 5908, Patents and Inventions, states that any invention conceived in the course of employment by a person under contract to the Department of Energy shall vest in the United States government, except when the administration waives such rights under provisions effecting the transfer of the invention to the inventor.

Chapter 20
Patent Reform?

Since the beginning of the United States patent system in 1790, there have been a number of amendments made to the patent laws and regulations. These amendments have been generally confined to patent office procedures and recognition of prior art residing in foreign patents. The International Patent Treaty has been instrumental in the advancement of technology by giving certain incentives to the advocates of cooperating nations and their inventors. However, the most important question respecting the degree of protection offered by a patent to the average inventor has not as yet been answered. Many inventors, with modest capitals to practice their inventions or to exploit through existing commercial sources, are losing out on benefits from their inventions.

To explain this situation, I will have to allude to solutions of more familiar problems. When a person owns a piece of furniture in his home, and one day when he returns home from a trip and finds the furniture stolen, he does not rush to sue just anyone. He notifies the police and they make a concerted search through available clues.

When a patent is appropriated by imprudent persons or manufacturers with ample capital and the patented product is manufactured and sold, no police or federal government agents enter into the litigation of the patented invention. In particular, the recovery of the patented product should be facilitated, in this case, by the fact that the "thief" is known and identified by his business that is commercializing the patented product without regard to the owner of the patent.

No police and no federal agents take part in any attempt to recover the "stolen property". If the patent is considered to be the property of the patentee or those he assigns, why shouldn't the state government or federal government commit itself to the recovery of the property that has been converted from an abstract right to a tangible product of monetary value? Does a patent really protect an invention from "vandals," thieves, and other criminals who believe the patentee cannot hurt them or their business because of his lack of funds to prosecute them in court? At the present time, the only recourse the patentee has is to sue the violator of his patent right in a federal court. If the patentee does not have the money to pursue such a course, there is nothing that can be done to arbitrate the dispute.

To initiate a court action, the patentee has to know whether his patent is valid or not. In order to determine this, he has to take the matter to the federal court to adjudicate his patent. If he thinks his patent is valid, he could bring an infringement suit against the infringer. Because taking a patent matter to a federal court is costly and generally outside the financial capability of the average inventor, his hands are tied and he can do nothing about it.

SHOULD PATENTS BE ABOLISHED?

Some who argue for the abolishment of the patent system suggest that patents often fail to reward the true inventor. Giant corporations would not favor change because they can afford to defend their patents against an infringer by hiring competent, high-cost attorneys to fight their case in court. The average inventor cannot financially bear the cost of a court action and therefore will lose out to a financially strong infringer. If an infringer also lacks the financial capability to fight a patentee in court, he will usually stop making and selling the infringed product for fear of a legal suit against him (and especially if he does not know the financial position of the patentee).

An infringer will frequently employ the services of a patent attorney. He will suggest that the patent is invalid by introducing a number of reasons not as yet proven in court. This often discourages the patent owner who usually gives up in dispair. If the patent is invalid, as alleged by the patent attorney, why was it granted in the first place? The patentee should direct the question of invalidity, suggested by the infringer's patent attorney, to the attention of the patent solicitor in the patent office. He can clarify the problem. The patent solicitor grants registration to an attorney who intends to practice before the patent office. Usually the background experience and character of an attorney is investigated before a registration is granted. Therefore, it is assumed that what the patent attorney declares should be authoritative. On the other hand, the patent office claims that any patent that has been granted by the office is assumed to be valid.

SHOULD PATENTS BE SUPPRESSED?

Some critics allege that certain companies take out patents on their developments and then shelve them to eliminate competition. Because patents are open to inspection by the public, it does not seem that such an expedient would suppress any progress of technology in the field in which the patent is granted. This type of action, if taken by a company, would not seem to effect anyone except the company. After spending time and money, it would seem unwise not to commercialize the product. The possibility does remain that the patentee does not consider the invention important enough to spend additional funds to set up a production line on a product he contemplates will not be profitable.

PATENTABILITY STANDARDS

In the remainder of this chapter, I will present suggestions and proposals for updating the administration of patents and patent regulations to make patent laws more equitable for the average inventor.

In accordance with the present patent laws, a person is considered not to be infringing on a patent if he infringes only on a single part of a patented invention. The concept behind this is that the entire patented invention has to be infringed before an infringement is committed and a suit to defend the patent can be valid (all other factors being equal). This notion not only is illogical in its entirety, but it is also unequitable. After having seen the entire structure of an invention illustrated in a patent, it would be relatively simple to substitute some vital part of the invention and make and sell the substituted invention that violates the original inventor's patent rights. This action actually amounts to using one species of the basic invention that has not been shown in the patentee's drawing or stated in his specification.

This is where the earlier instructions in this book come in. An inventor should preferably include in his first patent application, or in later applications, every feature that is possible in constructing his invention. Those so-called features are referred to as *species* of the invention. The inventor should be censured for his failure to take time and include in his application all possible ways of constructing his invention under the present patent laws. It would appear that minor improvements on the basic invention should belong to the first inventor and that any infringement on any single part of a patented invention should be considered as trespassing the patentee's property rights.

RAISING THE QUALIFICATIONS OF EXAMINERS

If the standards of the patent system are elevated, then it follows that the qualifications of the examiners should also be raised. The examiners should be law school graduates with at least one year of experience at a reputable patent law firm. They should also have a degree in one of the technical fields such as electronics, electrical engineering, chemistry, aeronautics, laser technology, metallurgy, mechanical engineering or other technical fields in order to avoid rejection of a patent application they do not understand on the grounds that "a dependent claim is a generic or species claim," as occurred in a recent examiner's documented action. Special, short weekly classes should be conducted at the patent office by more experienced and seasoned examiners in order to keep the lower-level examiners' knowledge up to date in current developments and principles and practice of patent examination.

RAISING THE SALARIES OF EXAMINERS

Because the qualifications of examiners should be raised, the salaries offered to examiners should also be increased in accordance with the grade of the examiner. Examiners should be paid more because their work involves, at times, high technology and their decision can lead to the advancement or failure of the technology in a given field. This often depends on their knowledge, education, and experience. At present, certain of the examiners qualifications do not measure up to the patent office reputation (although the U.S. patent office might be the best in the world). High salaries would attract high-caliber professionals to the patent office.

INCREASING THE NUMBER OF EXAMINERS

When the salaries of the examiners are comparable with those in private industry and educational institutions, more persons in both technical and legal disciplines will join the patent office force. At present, the patent application cases are slow and the situation will probably not improve in the near future. There is more work than the personnel can handle. It takes several years for a simple patent application to become a patent. The longer an application remains in the patent office the higher the cost of processing of the case. The filing and the final fees paid to the patent office cannot support the work load of the employees.

NORMALIZING THE PATENT OFFICE FEES

With improved patent office standards, highly professional personnel, and their increased salaries, the patent fees would have to be increased to compensate for improved activity at the patent office.

Public Law PL 96-517 is intended to bring in more money to the patent office. However, it will prove detrimental to the average inventor, advancement of technology, and the income to the patent office in the long run. Several movements are in the making with respect to boycotting the patent system. Members of various inventor's societies are working hard to effect a strike against patenting inventions. Is striking the remedy to this confusion? Will the new law nullify the inventor's incentives to invent? Would such a situation deprive the people of the United States of new technology that the inventions usually generate? These are serious questions and require adequate answers.

The new law, PL96-517, requires that maintenance fees be paid over a period of 12 years to keep a granted patent alive for 17 years. This scheme imitates the patent laws of foreign countries. The law stipulates the following:

First Maintenance Fee: $400 3½ years after grant of patent.
Second Maintenance Fee: $800 7½ years after grant of patent.
Third Maintenance Fee: $1,200 11½ years after grant of patent.

There is another important point that has been overlooked in drawing up the new patent law on patent fees. If our lawmakers are going to imitate, and have imitated partially, foreign patent laws, they should imitate the law in its entirety. While other countries do impose a maintenance fee upon the inventor, they also protect the inventor's patent against the infringer. The government that granted the patent will arbitrate disputes. Why don't we have the rest of the law as practiced in some foreign countries? The government should protect the inventor against the infringer at no cost to the inventor. Such a law would probably be received with high favor by many inventors because the inventor would feel secure that his patent is protected by his government.

The maintenance fee would amount to $1400 per patent, plus the filing and final fees, for a grand total of $1700. If the inventor has to pay a patent attorney's fee, the fee would be $1200 and up for a good patent attorney. The patent fees and the patent attorney's fee would be approximately $3000 for the complete patent.

If the inventor prepared his own patent application, he would save $1200 or more, but he would still have to pay $1700 or more. This is too high a price to pay for the average inventor. It is sometimes said that if the inventor had that much money to spare for a patent he would not have been an inventor in the first place because an average inventor is a working man with a family to support or a retired person trying to bolster his income with revenue from his inventions.

The revenue to the federal government from tax generated by the production and sale of inventions will be curtailed. Confusion will develop around the patent system because there will be persons who will not pay the patent maintenance fee after four years or eight years, and a purchaser of a patent will have no knowledge whether or not the patent is still in force. The money wasted or lost in this manner will be enormous. Unnecessary law suits will be initiated by those who have been infringed upon by inventors who will sell or license their inventions only to find themselves deeply involved in litigations. The lawyers will be involved in the prosecution of such cases in addition to infringement and adjudication of inventions. The inventor will be the one to lose his invention and money because of litigation. The public will also lose because development of inventions will be curtailed under these circumstances.

IS THERE A SOLUTION?

The new law, PL 96 517, should be abolished. In its place, another law should be established. The patent filing and final fees should be revised. For instance, the filing fee could be raised to $200 from the present $65 with the attendant fees remaining the same. The final fee could be $300 without additional cost for printing and drawing sheets. This would be a one-time fee so that when a patent is issued to the inventor there will be no doubt in the mind of any prospective buyer or licensee of the patent whether or not the patent is still in force. The date of patent issue would be a guarantee for such buyers or licensees.

The processing of the patent application should be accelerated and the time for granting a simple application should not exceed one year. The more complex inventions, of course, will take longer to process because of longer library search for prior art and for studying the lengthy specification. Because the total fee would be higher than at present, this would compensate for the increased time of examination and processing of the application in the patent office.

The standards of patent law practice should also be raised. We have many competent, honest, and knowledgeable patent professionals. Many of them are employed by large corporations. We need more such attorneys to go around for the benefit of the individual inventors when they need a lawyer for infringement law suits at a low cost (if infringements are not abolished).

The practice of adjudicating the issued patents should be curbed or eliminated entirely. When a patent is issued to an inventor, it should be a valid patent. Any infringement of the patent should be fought by both the patent office and the inventor in order to curb infringements. Just as in the case of police coming to the aid of a person who has been robbed, the patent office should appoint a branch whose duty should be to stop infringers whose attorneys claim a patent is invalid. The attorney who questions the validity of a patent in this manner must first show just cause for his allegations. If he cannot introduce reasonable evidence to support his allegations, he should be suspended from practicing before the patent office.

If the patent office cannot maintain the validity of the patents issued by it, as at present, and a court action is paramount to the settlement of infringement grievances, then the time between filing the suit and a decision advanced by the court should be reduced to a minimum. Competent professionals in the field of the contested patent should act as jury in support of a judge's decision. Each patent case should be handled on the strength of evidence presented, and the final decision respecting one case should not establish a precedent to terminate another case on the same basis. No circumstances could be identical in patent cases.

When an error is made in the wording of a claim or it is found that a claim is defectively written and requires correction, the court should not declare the entire patent invalid while other claims are tenable. The practice now is to correct the defective claim before filing a suit against an infringer. If not corrected prior to court action, the case is immediately thrown out. More tolerance should be exercised in such cases. The benefit of any doubt should be given to the patentee or owner of the patent because the patent office, in granting such a defective claim, is also at fault.

The matter of unpatentability of an invention due to obviousness of a certain part that has not been found in prior art should be amended so that more emphasis could be directed to what the invention contributes to the art rather than basing the entire invention on the part the examiner has declared obvious. There are many inventions that appear to be simple after they have been constructed. In order to develop the invention to the point of a workable product, it required many hours of thinking, amending, and redesigning. In the absence of exhibiting an invention in a working condition, it can be very difficult for the average person to design and develop a product. To illustrate this point, the following anecdote is given.

At a dinner party, Columbus was being harrassed by guests who claimed they could have discovered America if they had been given the chance. Columbus then picked up an egg and asked the guests if anyone could stand it on one end. Everyone retorted that it was impossible. Columbus broke one end of the egg and stood it on the dinner table. We could have done that, they all shouted. But, you did not think of it, said Columbus.

This is a good example to give to the examiner when he cites a certain part of the invention to be obvious. It might not be obvious until the inventor has achieved an operable product. Accordingly, the term *obvious* should be used with great care and deliberation by the examiner or the judge presiding on an infringement case.

A suggestion that might be noteworthy is to make the government a partner of the patentee. A certain percentage (such as 20 percent) of every patent that is granted could be assigned to the government so that, in an infringement case, the government could carry on the suit at its cost without any cost to the inventor. In this manner, the number of infringements would be reduced considerably. At present, many inventors with limited finances are discouraged and pushed aside by infringers' patent attorneys.

ABOLISHING INFRINGEMENT SUITS

This is a delicate subject. Patent attorneys do not want infringement suits to be abolished because they would lose revenue. Infringement suits are more costly than patenting the most complex invention. Therefore, the more infringement suits are called for, the more the patent attorneys stand to benefit. Perhaps the best method to approach the problem of infringement would be to establish an arbitration committee at the federal level under the patent office. The patent office is responsible for issuing patents and having authority over the patent attorneys. It is not the fault of an inventor if the patent he procured from the patent office has certain flaws in it or is invalid. If the patent is declared invalid, then all costs should be defrayed by the patent office if infringement suits are going to be sustained.

An arbitration committee would first gather all information regarding the invention and the prior art, and the circumstances under which the examiner has allowed the patent. When the committee decided that the patent office acted within its authority and that it made a thorough search to establish facts affecting the validity of the new invention, the committee would arbitrate the validity of the patent. This committee would comprise technical and legal professionals competent to pass judgment on patents and avoidance of prior art. In such an arbitration, minor improvements over the prior art would be considered and the benefit would be given to the infringer if such minor improvements are really not inventions (not withstanding any matter of obviousness). The decision of the committee would be final; no further action could be taken in any federal or state court.

PATENTS: THE SUPREME PROTECTOR OF INVENTIONS

Many people believe that a patent is an instrument in the hands of patent attorneys, and that they can manipulate it any way they deem necessary to their best interest. Such a notion is not true because a patent is a grant by the federal government to the inventor to exclude others from making, using, and selling the patented product without the consent of the patentee or patent owner. No attorney can alter the character of protection that is offered by a patent. He can declare his opinion that a patent is invalid, but he cannot invalidate a patent.

A patent granted by the United States Patent and Trademark Office should have the supreme authority over any allegations or assertations posed on it by any person unauthorized in the granting of the patent. For instance, a gold dollar cannot be changed into a copper dollar by the most expert chemist in the world. A patent protects the patentee and it should do so under all circumstances.

In trying to invalidate a patent, a person is attempting to nullify the credibility of the United States Patent Office. A patent should be an inventor's insurance that no unauthorized person is justified to take it away from him for a period of 17 years. No person should be able to infringe on it without heavy penalty and reimbursement of financial losses to the inventor. This should be automatic and not by tne force of a legal court. Until we attain such a state of patent posture, there will be much conflict, much confusion, and considerable waste of time and money. The next move is on the part of the public to urge the Congress to amend the patent laws that will more aptly suit the inventor as well as the patent office administration. It is then that new technology will be on the move.

Index

Edited by Steven Bolt